Technology of Industrial Materials

TECHNOLOGY OF INDUSTRIAL MATERIALS

By

H. C. KAZANAS

*Professor, Industrial Education
University of Missouri
Columbia, Missouri*

ROY S. KLEIN

*Assistant Professor, Engineering and Technology
Western Michigan University
Kalamazoo, Michigan*

JOHN R. LINDBECK

*Professor, Industrial Education
Western Michigan University
Kalamazoo, Michigan*

GLENCOE PUBLISHING COMPANY
BENNETT & McKNIGHT DIVISION

1988 Printing

Copyright © 1979 by H.C. Kazanas, Roy S. Klein, and John R. Lindbeck

Published by Glencoe Publishing Company, a division of Macmillan, Inc.

All rights reserved. No part of this book shall be reproduced or transmitted in any form or by any means, electronic or mechanical, including photocopying, recording, or by any information storage or retrieval system, without written permission from the Publisher.

Send all inquiries to:
Glencoe Publishing Company
15319 Chatsworth Street
Mission Hills, California 91345

Printed in the United States of America

ISBN 0-02-665900-X
(Previously ISBN 0-87002-301-2)

4 5 6 7 8   91 90 89 88 87

TABLE OF CONTENTS

INTRODUCTION 7

Chapter 1: What Materials Can Do 9
MANUFACTURING REQUIREMENTS, 9. SERVICE REQUIREMENTS, 10. PROPERTIES, Mechanical Properties, 11. Characteristic Properties, 14. SELECTION OF MATERIALS, 19. The Materials Revolution, 20. Picture Essay, 21.

Chapter 2: The Nature of Materials 27
ATOMIC THEORY, 27. The Structure of an Atom, Isotopes, 28. Electrons, Periodic Table, Electron Shell Theory, 29. BONDING, 33. Chemical Reactivity, 34. Physical State, 36. Bonding of Inert Gases, 37. Atomic Bonding and Chemical Change, Covalent Bond, 38. Ionic Bond, 40. Metallic Bond, 41. CRYSTAL STRUCTURE, 43. Allotropic Changes, STRUCTURES OF INDUSTRIAL MATERIALS, Industrial Metals, 45. Ceramics, 56. Organics, 58. BEHAVIOR OF INDUSTRIAL MATERIALS, Behavior under Load, 60. Behavior under Applied Field, 67. Magnetism, 69. Effects of Applied Energy, 71. Property Sensitivity, 72. READING LIST, 73.

Chapter 3: Metallic Materials 75
FERROUS MATERIALS, 75. Production of Pig Iron, 76. Production of Cast Iron, 80. Production of Steel, 84. Steel Mill Products, 88. Types of Steel, Composition, Properties, and Uses, 94. NONFERROUS METALLIC MATERIALS, Aluminum, 109. Copper, 115. Lead, Tin and Zinc, 120. Magnesium, 121. Precious Metals, 124. Refractory Metals, Other Nonferrous Metals, HEAT-TREATING OF METALLIC MATERIALS, Equipment for Heat-Treating, 125. Heat-Treating of Ferrous Metals, 126. Heat-Treating of Nonferrous Metals, 130. PROCESSING OF METALLIC MATERIALS AND PRODUCTS, Metal Cutting, 131. Metal Forming, 138. Metal Fastening, 148. Metal Finishing, 151. Special Metal Processes, 153. CONCLUSION—SELECTION OF METALLIC MATERIALS, READING LIST, 155.

Chapter 4: Polymeric Materials 157
DEFINITION OF PLASTIC, HOW PLASTICS ARE MADE, 158. Additives for Plastics, GENERAL PROPERTIES OF PLASTICS, 160. General Mechanical Behavior—Visco-Elasticity, 161. General Characteristic Properties, 162. CLASSIFICATION OF PLASTICS, Classification by Composition and Processing, Classification by Properties and Structures, PRINCIPLES OF STRUCTURE-PROPERTY RELATIONSHIPS, 164. Molecular Architecture, 165. Molecular Weight, Molecular Packing, APPLYING STRUCTURE-PROPERTY RELATIONSHIPS, Amorphous Plastics, 172. Crystalline Plastics, 176. Semicrystalline Structure, 178. Network Structure, 179. PLASTIC PROCESSES, Injection Molding, 181. Extrusion, 182. Compression Molding, Transfer Molding, 184. Blow Molding, Thermoforming, 185. Laminating, 186. Reinforcing, 188. Rotational Molding, Calendaring, 189. Coating, Foaming, 190. Other Plastic Processes, 192. READING LIST, 193.

Contents

Chapter 5: Ceramic Materials **195**
CERAMIC STRUCTURE, 195. PYROMETRIC CONE EQUIVALENT, RAW MATERIALS, 197. Raw Material Processing, FINISHED CERAMIC MATERIALS, Structural Clay Products, 199. Ceramic Whiteware, 200. Refractories, 202. Glass, 203. Porcelain Enamels, 206. Abrasives, 207. Cement and Concrete, 208. CERAMIC MANUFACTURING PROCESSES, 211. GLASS MANUFACTURING PROCESSES, 215. READING LIST, 216.

Chapter 6: Wood **217**
STRUCTURE OF WOOD, 217. Softwood and Hardwood, 218. Porosity, 219, PROPERTIES OF WOOD, 220, Moisture Content, Specific Gravity, 221. Physical Properties, Thermal Properties, 222. Electrical Properties, 223. Mechanical Properties, 224. Factors Affecting the Strength of Wood, 225. WOOD STANDARDS AND CLASSIFICATIONS, Lumber, 226. Other Wood Forms, 229. PROCESSING WOOD MATERIALS, 233. Wood as an Industrial Material, 235. READING LIST, 239.

Chapter 7: Miscellaneous Materials **241**
FIBERS AND FABRICS, The Nature of Fibers, Fabric Manufacturing, 241. Finishing Manufactured Textiles, LEATHER, 244. Synthetic Leathers, 249. COMPOSITE MATERIALS, Fibrous Composite Materials, 251. Laminar Composite Materials, Particulate Composite Materials, 253. LUBRICANTS, Petroleum Lubricants, 258. Synthetic Lubricants, 260. Fatty or Fixed Oils, 261. Lubricating Greases, 263. Solid Lubricants, FUELS, 265. Gaseous Fuels, 267. Solid Fuels, 269. Liquid Fuels, 272. Special Fuels, 278. READING LIST, 279.

Chapter 8: Testing of Industrial Materials **281**
NATURE AND SCOPE OF MATERIALS TESTING, 281. FAILURE IN MATERIALS, 282. TYPES OF TESTS—DESTRUCTIVE, 283. Tensile Tests, 284. Compression Tests, 293. Bending Tests, 294. Impact Tests, 297. Fatigue Tests, 302. Creep, Torsion, and Shear Tests, 304. Hardness Tests, 305. TYPES OF TESTS—NONDESTRUCTIVE, Visual Examination Tests, 318. Radiographic Tests, 320. Ultrasonic Tests, Magnetic Analysis Tests, 323. Electrical Analysis Tests, 325. TESTS FOR MISCELLANEOUS MATERIALS, Testing Concrete, 326. Testing Lubricants, 327. Testing Gaseous Fuels, 332. Testing of Solid Fuels, 333. Testing of Liquid Fuels, 334. CONCLUSION, READING LIST, 337.

APPENDIX **339**

GLOSSARY **373**

INDEX **387**

INTRODUCTION

Technology of Industrial Materials is designed to introduce students to the important materials of industry and how they are processed. This book is intended for use in community/junior colleges and also in four-year colleges where it should be especially helpful to students who are preparing to become industrial education teachers. It will be valuable for the training of technicians at all levels, and for reference purposes.

The purpose of *Technology of Industrial Materials* is to provide the student with broad understandings and concepts of the nature, processing, application and testing of industrial materials. The book is unique in that it concentrates on the basic principles of materials, rather than being merely a compendium of facts about woods, metals, and plastics. The practical applications of these principles are amply explained and illustrated. Thus the student can grasp the reasons underlying industrial materials applications.

The plan of this book is to introduce the student to the basics of materials technology, then to examine specific classes of materials. The first two chapters describe the physical and chemical properties of materials and the interrelation between structures and properties of the various classes of materials. Modern theories of materials science, such as dislocations in metals and molecular architecture of plastics, have been presented in a way that will be understandable to students without a strong background in science or mathematics. Industrial applications of these theories are given frequently.

The succeeding five chapters examine each class of materials in depth. These include metals, plastics, ceramics, wood, and miscellaneous industrial materials. In each of these chapters, the processing, properties, structure, and application of the materials are presented. The final chapter deals with testing—how the properties of materials are measured.

Although complete coverage is impossible, the aim of the authors has been to inculcate in the student a basis for understanding, a starting point for future exploration, and an interest in materials technology which he will carry with him into his career.

The authors would like to express appreciation to the many industrial firms and governmental agencies which have so generously supplied photographs and other illustrative material. The reader should note that these firms and agencies are actively engaged in materials research. This active research is an indication of the great importance that industry attaches to advancing materials technology.

H. C. Kazanas
Roy S. Klein
John R. Lindbeck

Chapter 1: What Materials Can Do

Technology can be defined as the application of scientific knowledge to practical purposes. When scientific knowledge of materials is combined with modern industrial methods, the result can be termed *industrial materials technology*. This combination makes it possible to convert natural materials efficiently into products which meet the needs and wants of people. Thus it is largely through industrial materials technology that the products necessary for a high standard of living are available to millions of people in many nations.

MANUFACTURING REQUIREMENTS

When a manufacturing firm considers making a new or improved product, it must examine both the technical and economical feasibility of doing so. Industrial materials technology is important in both considerations. Some of the major technical requirements can be met only if the materials to be used are properly selected, so that the product will perform satisfactorily in service. The economics of production is determined, to a great extent, by the expenses of acquiring raw materials and turning them into the desired finished product. These costs are also closely related to the choice of materials. Thus a proper knowledge of the properties and uses of materials will determine a major part of the economic and technical aspects of the product's success.

For example, consider the selection of a material for a television chassis. The designer or engineer faces several industrial considerations. Because the television set is scheduled for mass production, the chassis material must be easily formed by conventional equipment. The cost of the raw material must be reasonable. Additionally, the material must be strong enough to retain its shape and dimensional accuracy while supporting the electronic components.

Within these limits several choices are possible. Many metals are inexpensive, easily formed in standard presses, and machinable. Plastics are also easily shaped, though different processes are employed. Therefore the designer must compare such materials as steel and aluminum with the wide range of commercial plastics. The cost of the dies and machining fixtures for each material must be considered and so must the costs and other requirements for processing each material.

Beyond the need for inexpensive production, additional properties may be required of the material. It may be desirable for the chassis to conduct electricity. To help keep the television cool, the chassis should conduct the heat generated by the electronic components. The magnetic properties of the material must be considered, to avoid interference with the electronic circuits. The chassis may also have to exhibit specific acoustical properties to prevent damping or distortion of the tone.

The designer must first accurately estimate which of these properties will be needed in service applications. Then he must investigate the characteristics of each

Technology of Industrial Materials

material being considered. When these requirements of *service* are added to those of *manufacturing*, many materials can be excluded. However, new ideas, such as composites linking different materials may also be examined.

The material selected for the chassis is not likely to be the least expensive, the strongest, or the best as regards electromagnetic properties. However, it will probably exhibit all the required properties to some degree. The designer can allow for deficiencies. For example, to strengthen the chassis, reinforcing ribs may be added. To minimize electromagnetic interference, certain electronic components can be shielded or insulated. Thus the final choice is a compromise between the demands of the application and the properties of materials. (Incidentally, for a television chassis, a metallic material such as aluminum or steel is usually chosen.)

This simplified example stresses economics and performance as the major criteria. These are indeed major factors because they determine whether the product can be priced reasonably and whether it will satisfy the consumer. Price savings help distribute the fruits of technology to a wide population, but contemporary technologists must also consider ecological and safety factors. For example, will the material be readily recycled or disposable in current waste-control systems? Will the design contain sharp edges that can injure a TV repairman? The technologist can test his design against the standards of the Underwriters Laboratory, a safety agency sponsored by the elecrical industry, to insure against electrical hazards for the consumer. Environmental problems and the toll of accidental deaths (over 100,000 per year in the United States) warn the technologist that his work is of crucial importance. The things that shape modern life must be produced from the best materials using the best methods to improve the quality of modern life.

SERVICE REQUIREMENTS

There are certain general requirements for the proper selection of a material for a specific application. Designers and users of materials for industrial and technical applications should consider the following points:

- *Integrity of Shape.* If the product is to bear a load, its shape must remain within the tolerances specified by the user. This limits the use of low-strength materials to nonstructural or non-load-bearing applications, to prevent failure from excessive shape change.
- *Strength.* The material must resist breaking under an applied load. Since the same material will respond differently to various types of loads, the exact loading conditions of each application must be determined.
- *Integrity of Structure.* The temperature and other environment of the product in use must not cause structural changes that will make the material fail.
- *Service Life.* The desired useful life of a product, including storage prior to use, is the service life. A proper selection of materials requires knowledge of the service life and frequency of use of the finished material system. A fuse, for example, is used only once but has an expected service life of years.
- *Special Functions.* A product may have to perform a special function or service in respect to electrical, thermal, optical, or magnetic effects. If so, the selection of materials is quite narrowly limited to

those that possess properties for the service condition.

- *Formability.* Materials must be transformed into the shapes shown by the designer's blueprint. Therefore the formability of a material, how it reacts to industrial shaping, is a major consideration. The material selected must be readily formable to the required shape, using available industrial processing techniques. Ease of forming is one of the major reasons for the growing use of plastic.
- *Cost.* The overall cost of a finished product can often be reduced substantially by making the proper selection of materials and processing techniques. Overall cost is a combination of raw materials cost and cost of forming.
- *Environmental Effects.* The balance of nature can be adversely affected by certain materials which man is now making and introducing into his environment. Designers and engineers must consider the ecological consequences of their decisions. Such problems as waste disposal and recycling of used materials present new challenges in the selection of materials.

PROPERTIES

For proper material selection, it is necessary to know how the various materials under consideration will behave in the specific conditions of their intended use. The behavior of a material is discussed in terms of *properties.* Properties are the responses of the material itself, independent of size and shape, to specific mechanical and environmental loads or influences. Although many properties can be measured, only those which pertain to industrial applications are discussed in this book. These are broadly classified as mechanical and characteristic properties.

Mechanical Properties

Mechanical properties are related to the response of a material to applied forces and loads. The response may be deformation (shape change) or fracture. Thus mechanical properties indicate the strength and shape integrity of a material. Deformation characteristics of a material are useful indicators of its formability during industrial mechanical forming operations, such as rolling. Therefore understanding the major concepts of mechanical properties is a fundamental of materials technology.

Mechanical properties are usually determined by conducting tests, further explained in Chapter 8, that determine the resistance of the material to breaking. Therefore most mechanical property data relates to mechanical strength. But there are many ways of applying mechanical loads to a material. Fig. 1-1. Each method of loading can cause the material to respond in a different manner. Therefore a material actually possesses several types of mechanical strength.

Tensile Strength

A tensile load is one that tends to pull a material apart. Resistance to such a load is defined as *tensile strength.* When increasing tensile loads are applied, most materials initially change shape very slightly and will return to their original shape when the load is removed. This response to loading, termed *elastic behavior,* is the range within which a material retains its integrity of shape. Above a certain load (the *yield strength*) plastic response occurs and the material permanently deforms. Actual failure (fracture) occurs at still higher loads. The ultimate tensile strength of common steel is 70,000 psi (pounds per square inch), but it loses

Technology of Industrial Materials

shape at loads above its yield strength of 35,000 psi.

The tensile test is the simplest and easiest mechanical test to perform. For this reason it is widely used. Unfortunately the availability of tensile data has led many people to the mistaken notion that a material with a high tensile strength will exhibit the same high resistance to failure from other types of loading. Each type of loading must be evaluated to determine the performance of a material.

Compressive Strength

A compressive load is one that tends to squash a material. Resistance to this type of load is *compressive strength*. The mechanics of applying a compressive load and accurately measuring its effect on the material under study is more complex than the tensile test. Therefore compression tests are run only on certain types of materials which are used in compressive loading—concrete, for example. As will be shown in Chapter 2, the tensile and compressive behaviors of metals are similar up to the yield point. Because many forming operations such as rolling and forging are primarily compressive, compression tests involving major shape changes in metals are done to examine formability of a material subjected to these operations.

Fatigue Strength

Tensile and compressive strengths measure resistance only to continuously applied loads. Resistance to a *cyclic load* —that is, one that varies in direction and/or magnitude—is called *fatigue strength*. Fig. 1-2. Materials tend to be lower in fatigue strength than in tensile strength. Failure under cyclic load does not generally occur rapidly; rather, the product fails after some satisfactory service life. Hence, the term "fatigue": the material, so to speak, becomes "tired" of carrying a load below its tensile strength.

Some estimates indicate that fatigue causes 90% of industrial material failures. Generally, vibrations and rotational motion produce fatigue failures. Therefore the automotive, aircraft, and machinery industries have been faced with this problem for some time. However, it is only comparatively recently that materials scientists and technologists have discovered many of the reasons for fatigue failures. With this knowledge technologists are now improving the fatigue strength of materials.

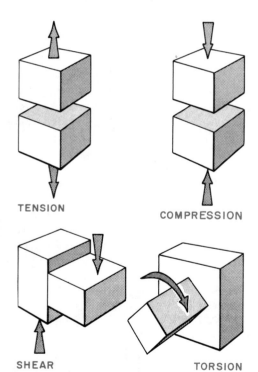

1-1. Mechanical properties. Various types of mechanical loads encountered in industrial practice. Another type, fatigue, is basically a result of repeated tension and compression loads.

measurement, usually either pounds per cubic foot or grams per cubic centimetre. The numerical ratio of this comparison is called the *specific gravity.*

Density is an important consideration when a material of low weight or controlled porosity is required. Steel and most other metals have a density of about 500 lbs. per cubic foot, whereas aluminum is about one-third this value, and plastics range within 65 to 100 lbs. per cubic foot. Products do not always exhibit the same density as their constituent materials. For example, metal castings contain random voids or porosity from processing errors. Filters and similar items require a material that has a controlled degree of porosity. However, in both cases a sample of sound material still exhibits its characteristic density. That is, when sound samples of the material used to produce the filter are tested, they will exhibit their characteristic density.

SELECTION OF MATERIALS

The final choice of material rests on the combination of properties, economics, and overall acceptability. As mentioned earlier, the exact requirements and conditions of the specific application must first be determined. Performance data from similar existing products provide the base point for the investigation. Stress analysis techniques can be used to calculate the types and magnitudes of mechanical loads of the new design. Models, mock-ups, and test runs simulate the range of service conditions encountered.

When the designer knows what a material will have to withstand in a given application, he must investigate the properties of different materials. By matching properties to service conditions, a group of acceptable materials can be found. Finally, the designer must take note of economic and ecological considerations to select the material best suited to his needs.

Let us follow the selection process for a wing panel located at the juncture of the wing and the body of an airplane. Fig. 1-9. First, the service requirements must be ascertained. By consulting the past performances of airplane wings, the engineer learns what general types of loads can be

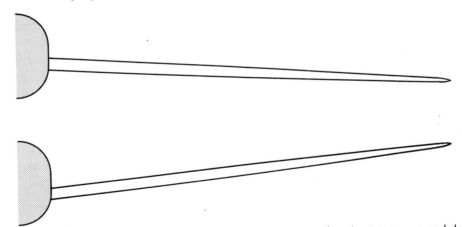

1-9. Fatigue load in wing section. Changes in air pressure cause the wing to move up and down repeatedly. When the wing is forced down, this causes a tension load at the top of the juncture with the fuselage, and a compression load at the bottom of the juncture. The loads are reversed when the wing is forced up. The repeated changes result in a fatigue load.

expected. Then, load analysis, stress calculations, and models are used to determine the specific tensile and compressive loads for the new plane.

Experience shows that other problems should be expected. For example, the continual air pressure changes in the atmosphere will cause the wing to move up and down, inducing a fatigue load. If the aircraft will operate near the sea, saltwater corrosion may occur. The wing may heat up from friction as it moves through the air at supersonic speeds. Flowing, oxidation, and expansion can occur at high temperatures. Although the wing member may not perform any specific electrical or optical functions, it can encounter stray electrical fields from guidance systems, or heat from absorbed light while parked on the runway. Finally, the wing might prove to be too heavy unless a low-density material is chosen.

After examining the requirements, the designer explores the appropriate property data for commercially available materials. He compares the responses of various materials to the specific applied loads. He also needs to know the chemical composition of the materials he is considering, and the manufacturing process, if any, which the materials have undergone. Often suppliers of materials will provide this information. Sometimes, however, sufficient data are not available for a material, thus necessitating a testing program in the laboratory.

The designer may find a suitable material, such as aluminum or plastic, already available to him. However, the design of a new product frequently involves requirements that are beyond the capacity of existing materials. The designer then requests the materials researcher to find new materials or combine existing ones in a new way to meet design requirements.

For example, titanium might replace aluminum because it is lighter, stronger, and resists high-temperature oxidation better. A composite of aluminum ribbing covered with thin sheets of high-strength plastic might also be considered.

The economic aspect of materials selection is primarily controlled by the cost of production. This includes purchasing raw material and converting it into the desired shape and quality. For example, the aluminum ribs may require extruding dies, and the plastic panels could be pressed on a forming die. Heat treatments, if any, and inspection costs for the finished wing must also be considered.

When the final choices of materials, processing techniques, and configuration are made, the engineers and designers must prepare instructions for the workers who will fabricate the wing. Material specifications, which give the specific chemical composition(s) of the material(s) and any required heat treatment, are issued. Draftsmen prepare engineering drawings that show the configuration and construction of the wing. Quality standards, specifying inspection of the raw materials and of the completed wing unit, are distributed. Thus the information required to proceed from technological concepts to a finished product can be distributed to the many people ultimately involved in producing the wing.

The Materials Revolution

The process of selecting a material was not always what it is today. Obviously, in primitive times choices were quite limited and therefore simple. Even fairly recently, materials selection was hampered by great gaps in knowledge. However, as you have probably realized from this chapter, the process of material selection as practiced in industry now is quite scientific. This

Chapter 1. What Materials Can Do

change has come about as a result of progress in two major fields. First, advances in understanding *service requirements* permit the accurate prediction of the types and magnitudes of loads to be encountered. Second, a great deal has been learned in recent years regarding the *internal structure of materials.*

As a result of these advances, designers are able to choose with great accuracy the one material which will best suit their needs. Moreover, particularly as a result of the new knowledge of materials structures, it is often possible to produce a material, when no ideal one is already available, to fill a specific need. This ability to control the structures of materials in order to produce desired properties has led to the so-called "materials revolution."

For example, the solid-state transistor represented a completely new type of material developed for a new application. Similarly, new materials can be made to perform existing functions more effectively or more economically, such as synthetic rubber and new clothing fibers. Moreover, existing materials can be modified to yield better performance. New high-strength and corrosion-resistant steel alloys used in building construction exemplify this.

Advances in materials technology have also made a major contribution to the materials revolution. The trial-and-error method, though still useful, is augmented by sophisticated testing techniques and inspection methods. Improved understanding and control of industrial processes result in uniform, reliable materials. Thus the modern materials technologist blends scientific insights with practical experience, leading to better products.

Picture Essay

An example of choosing a material for the wing panel of an airplane was given earlier. The following picture essay, Figs. 1-10–1-16, presents a similar example, concerned with choosing a material for making turbine blades for jet engines.

Materials used in jet engine turbine blades operate under very severe loading in a corrosive environment. Individual turbine blades are assembled into a turbine wheel which functions much like a water wheel. Hot gases from the combustion chamber are directed at the turbine wheels at temperatures over 2000° F. Fig. 1-10. The wheels rotate, producing mechanical power for equipment on the aircraft. In addition to high temperatures, turbine blades must withstand tensile loads, fatigue, creep, and corrosion. When selecting materials for turbine blades, all the conditions of loading must be considered. They are the specific design criteria and must be stated in terms of minumum strength and similar characteristics.

Fig. 1-10
COMBUSTION CHAMBER
⟶ HOT GASES ⟶

General Electric Company

TURBINES

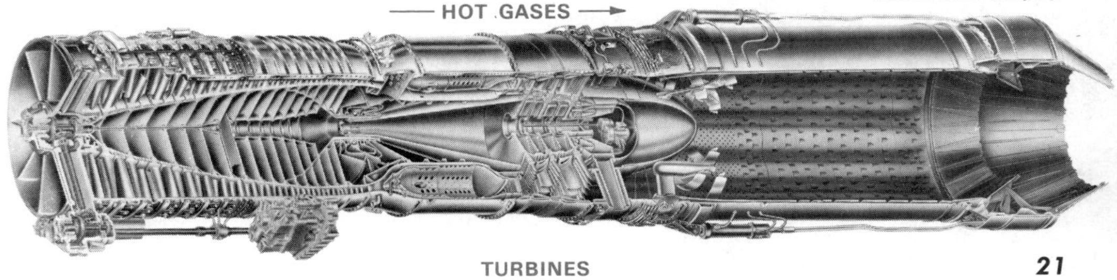

Fig. 1-11

Pratt & Whitney Aircraft

Fig. 1-12

Pratt & Whitney Aircraft

The material selected must exhibit a combination of properties which meet the design criteria. First, the designer usually considers commercially available materials. However, advances in turbine engine design frequently push loads beyond the capabilities of existing materials. Materials researchers must develop new materials, usually nickel or cobalt alloys. Fig. 1-11 shows a technician testing the stress rupture properties of several new alloys. An alloy which performs well in this test is then tested further. The effects of processing it industrially into a turbine blade are measured, using specimens from experimental blades. Fig. 1-12.

Chapter 2 — The Nature of Materials

Materials in infinite variety, natural and man-made, fill the world about us and affect our lives in countless ways. Men use these materials to give physical form to their ideas.

As explained in Chapter 1, the proper selection of materials is crucial for a satisfactory product, whether it be a unique object of art or a mass-produced hardware item. To simplify the selection process, materials can be classified in various groups which have many basic characteristics in common. For example, metals are known for conductivity and strength. Plastics are easy to form and resistant to corrosion. Wood is readily available and easily workable. (Subsequent chapters will examine the major classes of materials in detail.)

In the past few decades, the full range of scientific and technical knowledge has been brought to bear on understanding the nature, composition, and structure of materials. Under the broad topic of materials science, man is attempting to understand the materials that he has been using and to develop new ones to help improve the quality of human life. The purpose of this chapter is to relate the essentials of what science has learned about materials.

Prior to this century, man did not possess the scientific tools to examine matter closely. He had to rely primarily on direct, visual observation. For example, some ancient Greeks felt that the universe was built up from four basic materials—fire, air, water and earth. However, with the development of X-ray diffraction techniques and the electron microscope, man could begin to "look inside" matter. The emergence of sophisticated theories of matter in both physics and chemistry has brought man increasingly close to a full understanding of the basic nature of matter.

ATOMIC THEORY

Scientists feel that very small particles called atoms are the fundamental building blocks of all material. Since atoms cannot be broken down without loss of their chemical identity, the various kinds of atoms represent the most basic materials, the elements. (See Appendix A for alphabetical listing.) Only 92 different kinds of atoms are known to occur in nature. Some of the more common ones are oxygen, copper, iron, nitrogen, hydrogen, and carbon. (Besides the elements which occur in nature, man has produced at least 12 artificial elements through the techniques of nuclear physics, but these elements are unstable and have no common industrial applications.) The elements combine to form all other materials.

Atoms are constructed of various smaller bits of matter called *subatomic particles*. The three most important subatomic particles for the study of materials science are *protons, neutrons,* and *electrons.* Individual subatomic particles can be classified as to type but do not have a specific chemical identity as exhibited by atoms. A proton found in one atom will be the same as any other proton in any other atom.

The elements differ from each other only because the atoms of each element possess a specific and unique arrangement and number of subatomic particles. It is this difference which gives each element its characteristic chemical properties, such as reactivity or corrosion resistance. Thus it is clear that an understanding of chemical properties, which is essential to materials science, requires knowledge of atomic structure.

The Structure of an Atom

A single atom consists of a central portion or *nucleus* and the electrons that surround the nucleus. Fig. 2-1. The nucleus contains two types of particles: positively charged *protons* and electrically neutral *neutrons*. The number of protons in the nucleus determines what chemical element the atom represents; this number is called the *atomic number*. In the carbon atom illustrated in Fig. 2-1, for example, there are six protons in the nucleus. Therefore the atomic number of carbon is 6. Iron has an atomic number of 26; therefore every atom of iron has 26 protons in its nucleus. Fig. 2-2.

Compared to the other types of subatomic particles, the neutrons and protons are much heavier and approximately equal in weight. Therefore the materials scientist defines the total weight of an atom as the sum of the weights of its neutrons and protons. An arbitrary weight of one is assigned to each proton and neutron. Carbon, therefore, which has six protons and six neutrons in each atom, has an *atomic weight* of 12. Iron has 26 protons and 30 neutrons in the nucleus. Therefore the atomic weight of iron is 56.

Isotopes

However, when scientists determined the atomic weight of iron based on a large

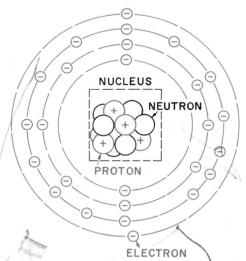

2-1. Structure of the carbon atom. The carbon atom contains 6 protons in the nucleus (shown in color) and 6 neutrons. Therefore carbon has an atomic number of 6 and an atomic weight of 12. To balance the electrically positive protons, 6 electrons, which have a negative charge, surround the nucleus.

2-2. Structure of the iron atom. The iron atom has an atomic number of 26 and an atomic weight of 56. There are 26 electrons surrounding the nucleus.

sample, they found that the atomic weight averaged about 55.85. Since the weight of each atom must be a whole number, they realized that a few iron atoms occurring in nature have an atomic weight of 55, not the normal 56. Atoms with such variations from the normal atomic weight are called *isotopes;* every naturally occurring element can exist in isotopic forms. The nucleus of an isotope contains more or fewer neutrons than the normal structure. Since the neutron has no charge and it does not affect the atomic number, the chemical identity and chemical properties of isotopes are the same as the common form of the element. But some isotopes are very unstable and radioactive. Scientists use these isotopes as chemical tracers. For example, some elements are known to collect in tumors. A sample of such an element containing radioactive isotopes can be injected into a patient's blood stream. A doctor can then locate the tumor, using a radioactivity detector.

Electrons

Surrounding the nucleus are the *electrons,* which carry an electrically negative charge. An atom in the free state must be electronically neutral. Therefore the number of electrons surrounding the nucleus must exactly equal the number of protons in the nucleus (*atomic number*). An electron weighs about 1/2,000 as much as a proton. Electrons move rapidly around the nucleus at velocities approaching the speed of light.

Periodic Table

If the elements are arranged by atomic number, and if they are also grouped by similar chemical and physical properties, a distinct order emerges. This grouping has been called the *periodic table* because it indicates the repetitive nature or periodicity of the common physical and chemical properties of the elements. Fig. 2-3. When the Russian chemist Dmitri Mendeleyev prepared the first periodic table about 100 years ago, many of the elements known today had not yet been discovered. However, by utilizing the repetitive pattern he was able to predict with great accuracy the general properties of the undiscovered elements. The periodic nature of the elements cannot be explained fully by studying the nucleus of the atoms; rather it requires careful examination of electron behavior as well.

Electron Shell Theory

The motion of the electrons about the nucleus suggests comparison with a miniature model of the solar system, with the nucleus as the sun and the electrons as planets in their orbits. However, the difference between this model and the actual behavior of subatomic particles is important in understanding the nature of materials.

Because electrons have such low mass and are moving at such high speeds, the common laws of physics that predict the behavior of much larger bodies do not apply. *Quantum mechanics*—a method of analysis based on energy levels, wavelike motion, and probability—was developed about fifty years ago, to describe mathematically the behavior of electrons. Although many areas of electron behavior are not yet known precisely, quantum mechanics enables us to make certain basic assumptions about electrons. The most unusual aspect of electron behavior (when compared with the normal behavior of matter) is that electrons act in a wavelike manner; much like light waves, they can be reflected or diffracted. This wavelike behavior makes it very difficult to estimate the locations or paths of electrons. It is

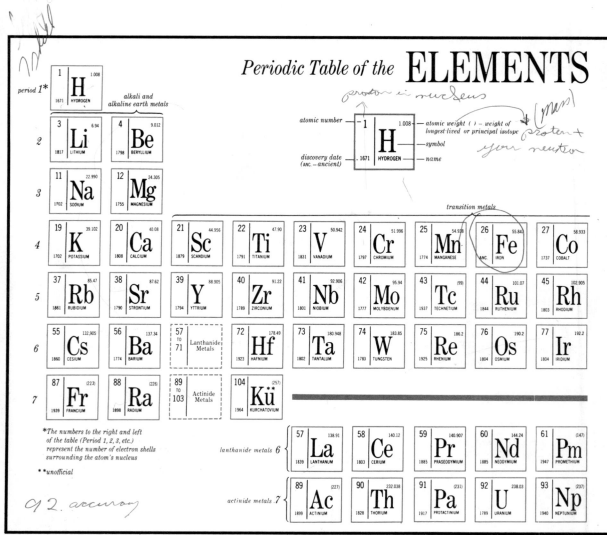

2-3. Periodic table. When the elements are arranged by atomic number, a repetitive or periodic pattern of properties appears. For example, the alkali metals (Li, Na, K) are very reactive whereas the inert gases (He, Ne, Ar) are not chemically reactive. The vertical rows are called groups or families. The horizontal rows are termed periods; they indicate the number of electron shells around the nucleus.

more accurate to talk about the *energy level* or energy content of an electron than about its location in space.

Since the location of an electron is not predictable, its orbits around the nucleus cannot be precisely compared with those of a planet around the sun. Rather than thinking of orbits measured in terms of distance from a central point, you might find it easier to think of orbits that represent energy levels—the more remote an orbit is from the nucleus, the greater the energy level it represents.

Also, unlike the solar system model, it is possible to have two electrons in the same orbit (rather than each in its own

Table 2-B
Relationship of bonding, structure, and properties.

Class	Material Name	Type of Bonding - Interatomic	Type of Bonding - Structural or Intermolecular	Boiling Point °F	Physical State at Room Temp. 75°F	Melting Point °F	Tensile Strength, PSI	Yield Strength, PSI	Characteristic Property Electrical Resistance Micro-ohm/cm§
Inert Gas	Argon	Van der Waals	—	−365	Gas	−370			
Inert Gas	Neon	Van der Waals	—	−435	Gas	−441			
Diatomic Gas	Oxygen (O_2)	Covalent	Van der Waals	−360	Gas	−428			
Diatomic Gas	Nitrogen (N_2)	Covalent	Van der Waals	−383	Gas	−410			
Organic	Methane (CH_4)	Covalent	Van der Waals	−368	Gas	−311			
Organic	Diamond (Carbon)	Covalent	Crystal	8500+	Solid	6000+			
Polymer Plastic	Natural Rubber Vulcanized	Covalent	Chemical		Solid	320*	3,000		10^{14}
Polymer Plastic	Polystyrene	Covalent	Liquid		Glassy	145*	7,000		10^{19}
Ceramic	Fosterite (Mg_2SiO_4)	Ionic	Crystal		Solid	3330	10,000		10^{17}
Ceramic	Glass (Window)	Ionic	Liquid		Glassy	1350*	10,000		10^{15}
Metal	Aluminum	Metallic	Crystal (FCC)	4380	Solid	1156	13,000	9,000	3
Metal	Steel (Iron-Carbon Alloy Heat-Treated)	Metallic	Crystal (BCC)	5500+	Solid	2675	200,000	150,000	6
Metal	1020 Steel Hot-Rolled	Metallic	Crystal (BCC)	5500+	Solid	2780	55,000	30,000	5
Metal	1020 Steel Cold-Rolled	Metallic	Crystal (BCC)	5500+	Solid	2780	61,000	51,000	6

*Softening Point §Micro-ohms per centimetre

outer-shell, high energy electrons—*the valence electrons*—that are responsible for the chemical reactivity of an atom. The chemical reactivity of a material depends on the relative ease with which it enters into electron reactions by gaining or losing valence electrons so that the outermost shell of the reacting atoms is filled. Chemical reaction will be examined more fully later.

If the groups of elements in the periodic table are examined, their chemical reactivity can be related to their electron structure. The *inert gases* represent a group of elements that are not chemically reactive with other materials. Referring back to Fig. 2-5 and Table 2-A, note that atoms of the inert gases have filled valence shells and therefore cannot react easily with other atoms. The most reactive elements are the alkali earths and the halogens. Fig. 2-3. These elements have either a shortage of one electron from a filled valence shell or they have a valence shell that contains only one electron. The reactivity of the elements in the remaining groups decreases as more than one or fewer than seven electrons appear in the valence shell.

Definition of a Metal

In materials technology, metals are defined by their properties—such as ductility, strength, and conductivity—and their structure. The chemist, however, considers that a *metal* is any element that will tend to give up or donate its valence electrons when bonded with another atom. The "chemical definition" is used in the periodic chart. Similarly, *nonmetals* are those elements that tend to gain or accept electrons when bonded to another atom. Since the atoms "attempt" to get a completed outer shell of eight electrons, metals are generally those elements with fewer than four valence electrons, whereas nonmetals have more than four electrons in the valence shell. Common metals, such as iron and copper, chemically react with nonmetals but *alloy* with other metals, as will be explained later.

In some elements, called the *transition metals,* the positions of the valence electrons are difficult to determine since the order of electron shell filling is not clearly understood. Also, some other elements with from three to five valence electrons do not exhibit a clear metallic or nonmetallic nature and can be either, depending on temperature.

Physical State

The physical state of an element—whether it is a solid, a liquid, or a gas at room temperature—is related to the type of bonding of the constituent atoms of the element. The physical state also gives an approximate indication of the strength of the bonds between the atoms of the elements when they are bonded together. A gas contains atoms that are very loosely bonded together and free to move quite independently of each other. In a liquid, the atoms are bonded together somewhat more tightly and cannot move independently. A solid represents the tightest bond. In the solid state, all the atoms are quite rigidly held in position by the surrounding atoms.

Some of the common behavior of the three states of matter can be discussed in terms of the degree of interatomic bonding that is present. A gas, because of the free movement of atoms mentioned above, will diffuse or spread throughout its container quite readily. For example, a girl wearing perfume can be detected at some distance, especially in a closed room. Perfume contains alcohol which will become a gas and diffuses or spreads

Chapter 2. The Nature of Materials

through the room, carrying the scent with it. However, when a liquid is placed in a container, it fills only the bottom of the container. The "tighter" bonding of the atoms in the liquid state prevents them from moving around freely and gives the liquid a greater density—more atoms per unit volume—than a gas. Tighter bonding of the liquid structure accounts for the viscosity of liquids—that is, their "thickness" or their resistance to flowing. Atoms of a solid are so rigidly bonded that the solid object maintains its shape unless enough force is applied to cause it to change; liquids and gases, by contrast, conform to the shapes of their containers. Obviously, the solid state is best for structural materials because materials in that state tend to maintain shape under an applied load.

Bonding of Inert Gases

The elements known as the *inert gases*—helium, argon, neon, and krypton—exemplify the relationship between properties and bonding. As was stated previously, these elements have a filled outer shell which represents a very stable or low-energy position. Therefore these atoms will have a very low attraction for each other and for atoms of other elements. Inert gases do not react readily with other substances because chemical reaction is based on a change in the electron structure of the reacting atoms. The electronic structure of the inert gases is too stable to change. Each atom can act quite independently of the others. This condition conforms to a very simplified definition of a gas—all the atoms that make up the gas can move around freely and do not have to maintain a rigid shape like the atoms in a solid or even the weaker molecular connections of a liquid. The boiling point of these elements—the temperature at which they will change from a liquid to a gas—is well below room temperature; thus they will be gases at the temperatures usually encountered in industry.

The chemical property of inertness makes these gases very useful in certain industrial processes in which it is important to prevent elements that are more chemically active from reacting with the workpiece and changing its composition. For example, a very common industrial process is inert gas welding. In this process a protective blanket of helium or argon is maintained around the weld to prevent chemical attack by air. Similarly, incandescent electric light bulbs are filled with neon or krypton to protect the hot filament from atmospheric attack.

Van der Waals Forces

It might be thought that atoms of inert gases would have no attraction for each other, since their valence shells are filled. In fact, however, a weak attraction does exist. This attraction has been named the van der Waals forces in honor of a Dutch physicist who helped explain it. Van der Waals forces result because the motion of the valence electrons in one atom influences valence electron motion in nearby atoms. For example, if most of the valence electrons in atom A (Fig. 2-6) moved to the right side of the atom, they would tend to repel the valence electrons in B and push them to the right.

Van der Waals forces can occur between any two atoms, not just inert gases. That is, whenever two atoms have valence electrons that move in harmony, a small attractive force exists between them. However, in the atoms of most elements these forces are masked by more powerful bonding forces. Even in the inert gases the van der Waals forces are very weak because the eight valence electrons tend to randomize rather than harmonize the orientation of their motion.

A unique characteristic of van der Waals forces is their ability to act at relatively long ranges—several atomic diameters. They are the only known long-range interatomic bonding forces. To convert a gas into a liquid, when the only bonds between the particles are the van der Waals forces, industrial processes use high pressures. The pressure can be thought of as actually pressing the atoms together so that the stronger bonding needed for liquefaction can take place.

Atomic Bonding and Chemical Change

Only the inert gases can occur in nature as essentially individual atoms loosely held together by the van der Waals forces. Atoms of all other elements, as mentioned earlier, exchange valence electrons to achieve the more stable condition of a filled valence shell. These elements, in the pure state, exhibit higher boiling points, indicating tighter bonds, and usually exist in combinations of two or more atoms (called molecules).

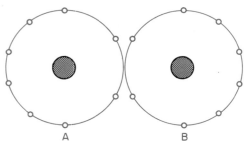

2-6. Van der Waals forces. These forces result from the electrons in two neighboring atoms acting in harmony. When a majority of the valence electrons in atom A move to the right, the valence electrons of B also move right. If the electrons in A shift closer to B, they repel the similarly charged B electrons to the left. Thus the valence electrons in both atoms tend to "cooperate," producing van der Waals forces.

When two or more atoms interact to alter their electronic structures, a *chemical change* occurs. Such changes also alter the energy levels of the atom involved—that is, the combinations formed through the chemical change have lower energy levels than did the atoms prior to the change.

Atoms form three common types of bonds—*covalent, ionic,* and *metallic.* Specific types of bonding can be associated with the various classes of industrial materials. In the following pages, these three bonds will be examined to help you understand their effect on the properties of materials.

Covalent Bond

Many common gases, such as hydrogen, oxygen, and nitrogen, are *diatomic*—that is, they form molecules consisting of two atoms bonded together. The chemical abbreviations for these diatomic gas molecules are H_2, O_2, and N_2.

The bond between the atoms of each molecule can perhaps be best explained by an example. Oxygen has six electrons in the outer shell. When two oxygen atoms combine, the result obviously is not a molecule with a filled outer shell of eight electrons. Rather, oxygen atoms share enough of their electrons to fill their respective valence shells. Each oxygen atom needs two valence electrons; therefore each shares two of its electrons with the other atom of its molecule, thus in effect giving each atom a filled valence shell. The two atoms are bonded quite strongly by the shared electrons. Interatomic bonding based on the sharing of valence electrons is called *covalent bonding*.

Not only are many gas molecules covalently bonded, but all *organic* compounds (those compounds usually based on carbon) are also covalently bonded. An example of the organic covalent bond is the

Chapter 2. The Nature of Materials

methane gas molecule, CH_4, illustrated in Fig. 2-7. The carbon atom has four electrons in the valence shell, each of which is shared with the valence electron of a hydrogen atom. This results in a stable, filled shell for all five atoms, binding them into a single molecule. Note that there is a definite shape to the methane molecule because the hydrogen atoms tend to repel each other and keep themselves spaced apart. The paired electrons are located between their parent atoms. The shape of organic molecules is a characteristic of the specific atoms that are bonded together.

It is possible to construct very large molecules through covalent organic bonding. The methane-type molecule can be expanded by substituting a carbon atom for one of the hydrogen atoms and then adding two more hydrogens to fill the valence shell of the carbon. Fig. 2-8. This process can be repeated to increase the size of the molecule. The molecules have a specific configuration. Gasoline and plastics are composed of molecules that have been greatly expanded, sometimes to as much as several thousand times the original size.

The high strength of the covalent bond can be illustrated more clearly when the

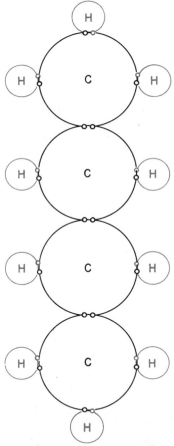

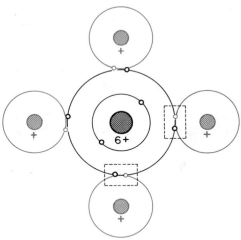

2-7. The covalent bond. A molecule of methane has been formed by the carbon atom sharing valence electrons with 4 hydrogen atoms. Each hydrogen atom shares its valence electron with the carbon atom. The shared electrons are shown in the dotted squares. The carbon atom in the methane molecule has a filled outer shell with 8 electrons, and each hydrogen has a filled 1s shell containing 2 electrons.

2-8. Large organic molecule. The covalent bond of carbon permits the development of large molecules, as shown in this simplified diagram of a polymer. Large organic molecules are the basis of living tissue, including wood, as well as of plastic and petroleum. Note that only the shared electrons are shown.

diamond structure of carbon is examined. Each carbon atom shares one valence electron with a neighboring carbon atom. Thus the entire piece of diamond can be pictured as one giant molecule. The diamond represents one of the hardest materials known and melts at a very high temperature. These physical properties attest to the high strength of the covalent bond.

On the other hand, many materials with covalent bonding have the physical properties of weaker bonds, such as low strength and low melting points. For example, the diatomic gases mentioned at the beginning of this section have boiling points only slightly above the inert gases, though each molecule of the gas is covalently bonded. This apparent discrepancy occurs because the forces between the molecules are not as strong as the covalent bonds between the atoms. The diatomic gases rely on van der Waals forces between their constituent molecules, similar to the inert gases. Many organic materials, such as polymers, including plastics, consist of aggregates of large numbers of covalently bonded molecules. But many properties of such an aggregate depend on the intermolecular forces, which will be examined in a later section.

Ionic Bond

There is another method of forming chemical compounds based on an interchange of electrons rather than on electron sharing. An *electron donor* atom can combine with an *electron acceptor* atom to form a molecule by an interchange of a valence electron. For example, a chlorine atom (atomic number 17) needs only one more electron to have its outer shell filled. If a chlorine atom is brought into the vicinity of a sodium atom (atomic number 11) which has only one electron in the outermost shell, the sodium would give up its outermost electron to the chlorine. Fig. 2-9. This would completely fill the outer shell of the chlorine atom and leave a filled outer shell in the sodium atom. As previously stated, atoms with completed outer shells are stable.

By definition, an atom must be electrically neutral. After the chlorine picks up the electron of the sodium atom, its number of negative charges exceeds the number of positive charges by one, resulting in a net charge for the atom of *negative one*. Similarly, the sodium has a net charge of *positive one* because it has given up one electron but still has 11 protons in the nucleus. Atoms that are not electrically neutral are called *ions*. The electrical attraction of the ions holds these two atoms together, as a molecule, very tightly, in what is called the *ionic bond*. (The molecule is electrically neutral after ionic bonding has occurred.) In the example just described, the molecule formed is

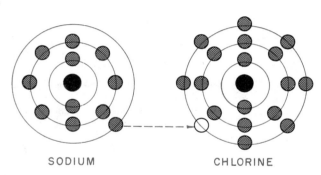

2-9. The ionic bond. A molecule of sodium chloride is formed when the sodium atom (left) donates its single valence electron to the valence shell of the chlorine atom. The chlorine atom has 7 electrons in the valence shell. After the electron interaction, each ion has a filled valence shell.

SODIUM   CHLORINE

Chapter 2. The Nature of Materials

sodium chloride. Using chemical nomenclature, chlorine changes to chloride, indicating that the sodium and chlorine are bonded together, not free elements. Several atoms can be ionically bonded into a molecule as shown in Fig. 2-10.

Large numbers of ionically bonded pairs of atoms are found arranged so that the positively charged ions are surrounded by negative ions of other pairs. The electrical attraction will bind the pairs together. Similarly, the negative ions are surrounded by positive charges. Fig. 2-11 shows the resulting arrangement of ions, alternating with respect to charge. Each ion is held in place by the balance between the electronic attractions of the oppositely charged ions surrounding it. Regularity of spacing results from this balance of attractive forces, producing a *lattice structure* or *crystal structure*.

Most common chemical compounds are bonded ionically. This includes table salt (sodium chloride) and many of the ceramic components of sand, such as aluminum oxide and sodium oxide. Ionically bonded materials tend to be hard and have high melting points because the lattice bonding is very strong between all their atoms. The high melting point of ionically bonded materials and the relative chemical inertness of this type of bond make many ionic compounds useful in ceramics and refractories.

Metallic Bond

Thus far, two types of bonding based on the localized interchange of valence electrons between a small number of atoms have been presented. Metals typically represent a different type of bond because all atoms in the structure (or aggregate) have a similar valence and cannot directly exchange electrons to form

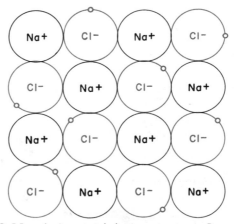

2-11. Ionic crystal lattice structure. Representation of a crystal lattice structure produced by ionic bonding. Nearest neighbor ions are at the same distance, and the structure is repetitive. The electron donated by the sodium to each chlorine atom is shown.

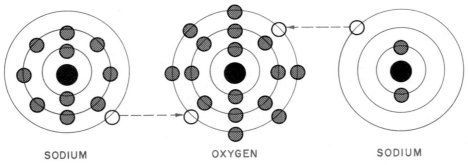

SODIUM   OXYGEN   SODIUM

2-10. Molecule of sodium oxide. Each atom of sodium donates its valence electron to complete the valence shell of the oxygen. All three atoms are now bonded together ionically.

completed outer shells. The *metallic bond* is based on the ability of metal atoms to donate their valence electrons to a common pool of electrons, termed the *electron cloud*. The electron cloud can exist because the energy level of the valence electrons of the atoms in the aggregate splits into enough subshells of only slightly different levels so that all available electrons can be accommodated. The valence electrons in the cloud are not localized; they are relatively free to travel throughout the structure in random directions and at enormous velocity. The random movement of the electrons is possible because the energy differences are so slight that it is easy for an electron to change subshells and find an available position in a nearby atom. The randomness of the motion also equalizes the charge of the electrons in the structure. Fig. 2-12.

Metallic bonding results in a structure that consists of (1) positively charged ions with a complete outer shell and (2) a negatively charged electron cloud "moving" through the structure. The electron cloud has been called the "glue" or binding force of the metallic state. Because the cloud can accommodate any number of electrons by splitting the energy level it represents, a piece of metal can be made into any size or shape. (It will be shown later that the electron cloud also accounts for the electrical conductivity of metals.)

When metal atoms are brought into sufficiently close contact to form the metallic bond, valence electrons from each atom join the electron cloud. The charged particles remaining in each atom then are no longer in balance. The atoms thus become ions, and various attractive and repulsive forces begin to interact. The ions will repel each other because both, having given up negative charges, are now positively charged. The remaining outer electrons of one ion will also repel those of the other ions. However, these repelling forces are counterbalanced by the attraction between electrons (negative) of one ion and the core (positive) of another ion.

The exact magnitude of the attractive and repulsive forces is related to the distance between the ions. Fig. 2-13. At very small separations, the repulsive force predominates. At large spacings, the attractive force is greater. However, there is an intermediate spacing at which these two forces are balanced and the energy is at a minimum. This is the most stable configuration for the atoms to assume. Once two atoms have been metallically bonded (thus becoming ions) it is difficult to change their spacing from the stable or equilibrium distance. Compressing them would increase the repulsive force between the ions, which would tend to restore the stable spacing. Similarly, the attempt to enlarge the spacing would be resisted by attractive forces between the ions. It is the action of these forces which accounts for the high tensile and compressive strengths of crystalline materials.

The exact distance between any two atoms is related to their orientation and

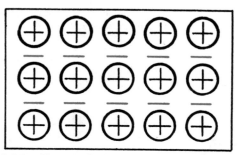

2-12. The metallic bond. The metal atoms become positive ions by donating valence electrons to the electron cloud. The negative charge of the electron cloud, indicated by the minus sign, is uniformly distributed throughout the structure and serves to bond the positive ions together.

the number of charged particles. The distance between any pair of like atoms—atoms of the same element—will be the same; for each element the distance will be different. The characteristic *lattice parameter* or *interatomic distance* of elements is the basis of identifying them by *X-ray diffraction analysis*.

If the concept of a specific, repetitive interatomic distance is extended into three dimensions, a lattice-type structure made up of ions is developed. The metallic lattice is geometrically similar to the ionic lattice but differs in charge distribution because all metal ions have the same charge. In a model of a metallic lattice, metallic ions could be represented by balls or spheres, since the outer electrons have been removed and are in the electron cloud. The ions are fixed in specific positions by the balance of electronic forces with the neighboring ions.

CRYSTAL STRUCTURE

When a large number of ions are arranged in a repetitive pattern to correspond to a lattice, the result is a crystal or grain. In a metal, the atoms nearest to each other attract each other through the free interchange of valence electrons in the electron cloud.

The arrangement of atoms in a crystal can be compared with the arrangement of balls in a box. Assume that the balls are all of the same size, that each ball is equidistant from its neighbors, and that there are no voids (holes where a ball is missing). Given these conditions, there are several ways to pack the balls into the box. Similarly, there are several arrangements in which lattices can be constructed. Through the use of X-ray diffraction analysis, materials technologists have determined that metals will form in one of three crystal lattice arrangements: body-centered cubic (BCC), face-centered cubic (FCC), and hexagonal close-packed (HCP). Most metals in industrial use are either FCC or BCC; therefore these structures will be examined.

Every crystal lattice has its own *unit cell*. You will remember that an atom can be

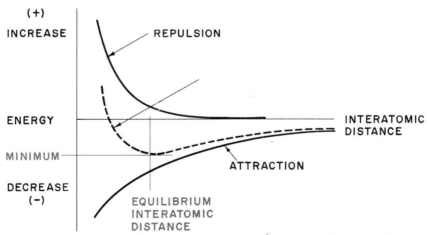

2-13. Interatomic spacing in a lattice. As atoms are brought together, they both attract and repel each other. The dotted curve indicates the sum of these forces. At a specific distance, the sum is a minimum, indicating a stable, low-energy position. This is shown in color and indicates the equilibrium spacing between the atoms in the lattice.

defined as the smallest particle of an element which exhibits all the characteristics of the element. In a similar way, the unit cell is the simplest model (with the fewest atoms) of a crystal lattice which exhibits all the characteristics of the lattice. The reason that some crystal lattices are called body centered is that their unit cell is a cube with an atom at the center of the body of the cube. In other words, an atom is centered *inside* the cube. Fig. 2-14.

For comparison, consider a box that is filled with balls in this manner: The first layer of balls is placed on the bottom. The next layer is stacked so that each ball rests in the hollows between groups of four balls in the first layer. Succeeding layers are added in the same way. This is basically the structure of the BCC crystal. If the balls were atoms in a unit cell, and if the cell were replicated, the result would be a BCC crystal. Some metals that exhibit a BCC crystal structure are iron (steel), chromium, and molybdenum.

In the face-centered cubic structure, each atom in the second layer rests on a hollow between two first-layer atoms and has its center in the same vertical plane as the two atoms on which it is resting. Fig. 2-15. The unit cell for the FCC structure has the appearance of a cube that has an atom *at the center of each face.* Aluminum, copper, gold, and silver exhibit an FCC crystal structure. Copper crystal, for example, has an FCC structure in which the distance between nearest atoms is 3.6153 angstroms. (An angstrom is an extremely minute unit of measurement based on wavelengths.)

The essential characteristics of the metallic class of materials is the combination of the crystal structure and the electron cloud. This unique structure is responsible for the properties that are associated with metals. Therefore the number of elements that can form metals is limited to those atoms which donate electrons. Such elements usually exhibit a valence of three or less; however not all electron donor elements can form the metallic structure. Note that chemists define metallic state by the valence, whereas in materials technology, the *structure* is the basis for classification.

As previously mentioned, through ionic bonding some nonmetallic materials have crystal structures. These materials, such as sodium chloride and magnesium oxide, are similar to metals in that they have characteristics of spacing and repetitive

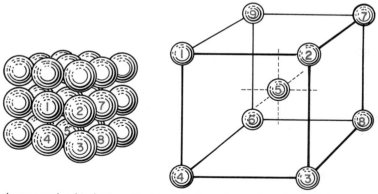

2-14. Body-centered cubic lattice. Stacking of atoms to produce the BCC lattice structure. The numbered atoms represent the unit cell of the BCC structure.

array which resemble those of metal crystals; however, the nonmetallic crystals do not exhibit an electron cloud. Organically bonded large molecules can also exhibit a crystal-like structure, which is discussed more fully in Chapter 4.

Allotropic Changes

Certain elements can exist in two or more crystal structures. The alternative structures are termed *allotropes* and are dependent upon temperature and pressure. For example, pure iron exists in a BCC structure at room temperature but undergoes an allotropic change to FCC if the temperature is raised above 1670° F. Carbon undergoes three common allotropic changes. If exposed to high temperature and pressure, the covalently bonded crystal structure of a diamond results. Coal, with no definite structural arrangement, occurs at lower temperatures and pressures. The allotropic form of carbon called *graphite* has a structure of sheets of covalently bonded carbon atoms held together by weak bonds so that the sheets can flow over each other and act as a lubricant. Allotropes can also be called *polymorphs*.

Chapter 2. The Nature of Materials

STRUCTURES OF INDUSTRIAL MATERIALS

It is necessary to understand the concepts of bonding small numbers of atoms. However, certain important differences arise when the concepts are applied to large and varied aggregates of atoms, as when working with most industrial materials. Such materials are rarely chemically pure. They are generally, and intentionally, a combination of elements and materials, selected to give the desired mechanical or chemical properties for a specific application. The most commonly used industrial materials are metals, ceramics, and organics (plastics and wood polymers). Each type will be examined to determine representative structures and the generalized properties to be expected.

Industrial Metals

The previous discussions were primarily concerned with single crystals of pure metals. Certain applications in the electronics field require materials of this type. However, the vast majority of metallic materials used in industrial applications are composed of several types of atoms mixed together. Such materials are termed

 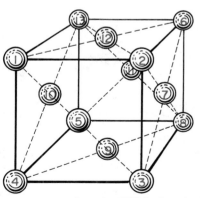

2-15. Face-centered cubic lattice. Stacking of atoms to produce the FCC lattice structure. The numbered atoms represent the unit cell of the FCC structure.

Technology of Industrial Materials

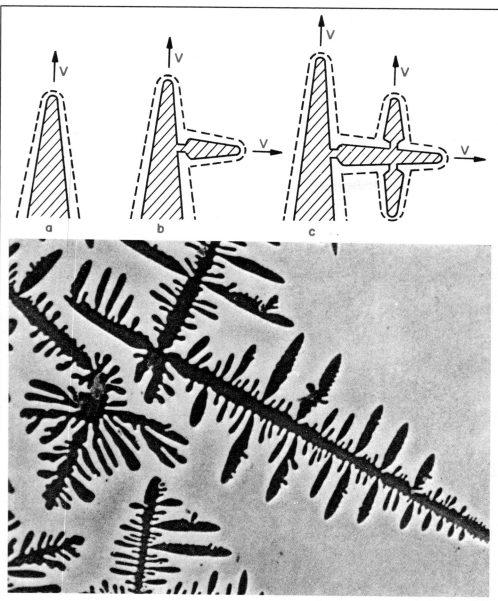

2-16. Dendrite formation. In the top view, the direction of growth is shown by arrows marked V. Each nucleus starts growing along a specific crystallographic direction (a). Then the original "trunk" sends off "branches" (b) which in turn branch out (c). In the bottom view, the resulting treelike or dendrite growth of a solid metallic phase within a liquid metal bath has been simulated by using plastics that solidify by a mechanism similar to that shown in the drawings. Upon solidification, each dendrite has produced a grain because the remaining liquid fills in the spaces between the branches.

alloys. They are *polycrystalline,* which means they contain many crystals. The use of polycrystalline alloys permits the development of the wide range of properties exhibited by metals.

Grain Structure

Metallic parts used in industry consist of many individual crystals that are bonded together rather than a single large crystal. It is customary to refer to the polycrystalline nature of metals as the *grain structure.* The terms *crystal* and *grain* may be used interchangeably in a technical discussion of metals.

Metals go through certain changes called *phase transformations.* The allotropic structural changes mentioned earlier are phase transformations. So are solidification, melting, and other changes that take place in the processing of metal components. The grain structure of metals is altered by these transformations. If the structure is altered in such a way that the crystals of the new phase are grown or produced from the pre-existing structure, then a phase transformation is said to have taken place.

When a grain structure begins to change—that is, when a phase transformation begins to take place—the part of the new grain structure which forms first is called the *nucleus.* (This should not be confused with the nucleus of an atom.) Nucleus in this sense means an extremely tiny particle which serves as a base for further growth or spread of the crystal structure. As mentioned, the passage of a metal from one phase to another is marked by a change in the grain structure. However, the change is not instantaneous—that is, nuclei characteristic of one phase may be present in material which is mostly in another phase. However, as the change accelerates, atoms which had

Chapter 2. The Nature of Materials

been part of the old structure align themselves with nuclei of the new structure until the change is complete. Usually each nucleus grows in a pattern that resembles a tree, as shown in Fig. 2-16, leading to the name *dendrite* (Greek for treelike).

The development of a polycrystalline structure during the solidification of a metal is depicted in Fig. 2-17. The molten metal is poured into a mold at room temperature. The mold chills the metal below the melting point. A number of nuclei form depending on the amount of supercooling (rapid cooling below the phase-change temperature). Metal atoms from the melt add on to the nuclei of the solid until all the metal becomes solidified. Each stable nucleus formed has developed into a grain in the final structure of the solid metal. Each grain of the solid has the same crystal structure, BCC in this example, but each grain is oriented somewhat differently from its neighbors. Fig.

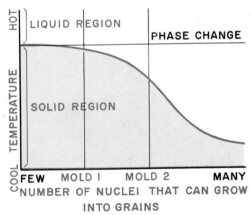

2-17a. Polycrystalline grain size. The average grain size depends on the number of nuclei which form when solidification begins. This chart indicates what has happened to two molds of molten metal which were at the same initial temperature when cast. In Mold 2 the liquid was supercooled—chilled below the solidification temperature—to a greater extent than in Mold 1. (The amount of supercooling is shown in color.)

Technology of Industrial Materials

2-17b shows the difference in orientation. Note that the nuclei are not aligned uniformly, which results in the formation of grain boundaries.

When the grains are growing together, the atoms in the boundaries between two grains must take locations that are not equilibrium positions for either grain. Fig. 2-18. The grain-boundary atoms will have a higher energy level than interior atoms, which are in equilibrium positions relative to their nearest neighbors. Grain-boundary atoms will react at a different rate than the interior atoms because of this energy differential. This fact can be used to determine the average grain size

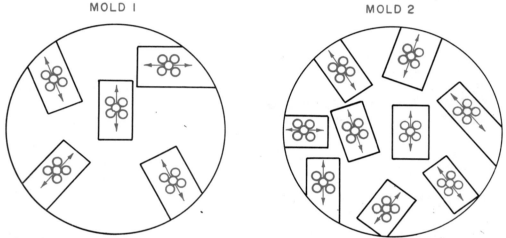

2-17b. Nucleus formation. More nuclei have formed in Mold 2 than in Mold 1. This is because of the greater supercooling of Mold 2, as shown in Fig. 2-17a. Each nucleus has a specific crystallographic orientation and starts to grow in a dendrite pattern.

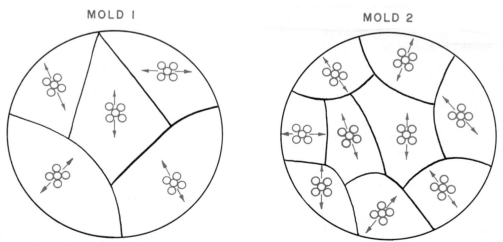

2-17c. Solidified metal. Each nucleus has grown into a grain of solid metal. Mold 2 has a smaller average grain size than Mold 1.

Chapter 2. The Nature of Materials

of a metal part. By applying a suitable chemical reagent to the metal surface and viewing under a microscope, the grain becomes visible so the size can be visually determined. Fig. 2-19. The variations in shading of the individual crystals are caused by the differences in orientation between the grains discussed previously.

Grain size is important because it influences the strength and formability of metals. At room temperature, a fine-grained metal part—that is, one with a small average grain size—is generally stronger and tougher than a coarse-grained part of the same metal. This is attributed to two causes. First, grain boundaries are stronger than the grains themselves, so that a fine-grained part has, on the average, more boundaries. Second, any failure within a grain will be smaller and less likely to spread if the grain itself is small. A simple analogy is the higher resistance to cracking of a brick wall (fine

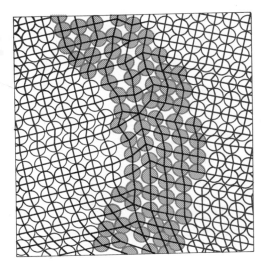

2-18. Grain boundary structure. As the crystals grow toward each other, the atoms at the junctures must adopt positions that are not equilibrium positions for either grain. These atoms at the grain boundaries are high in energy because they are not in equilibrium positions.

2-19. Grain size and orientation. Variations in grain size are easily observed when comparing these two highly magnified metal samples. Note also the pronounced difference in grain orientation. It is this which causes differences in light reflected from the surface. When polished, etched, and viewed (usually under a microscope) the light and dark grains indicate the differences in orientation.

Ford Motor Company

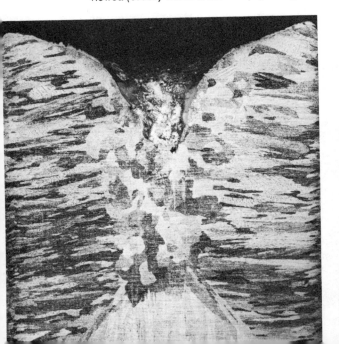
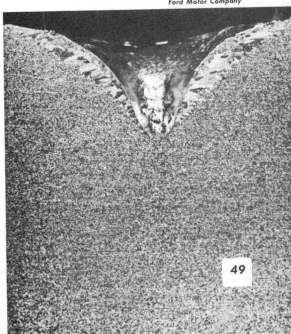

Technology of Industrial Materials

grained) as compared to a block wall (coarse grained).

To produce a fine-grained metal, chemical composition and processing techniques are controlled. For example, in aluminum making, certain alloys are added that help form nucleating sites for growth. Rapid cooling through a phase change results in a fine-grained structure. Industrially, this can be achieved by quenching during heat-treating or by casting into a material with a high heat conductivity, such as an iron mold instead of refractory sand. As illustrated in Fig. 2-17, rapid cooling would result in greater undercooling, (or supercooling), more nuclei, and thus a finer grain size.

Working of a material can result in changes in grain size and orientation. When a metal is rolled or formed, the grains can become elongated in the direction of the forming operation. Fig. 2-20. This structure is characteristic of cold-worked or cold-rolled steel, copper, and aluminum. If the distorted grains are heated for a specific period of time and above a minimum temperature, the new, undistorted grains will grow from the worked structure. (Both the time and the temperature are empirically determined for each metal.) This process, termed *recrystallization,* results in a soft, ductile state for the metal.

Alloys

An alloy is a metal which contains lesser amounts of other elements, either metallic or nonmetallic. The alloying elements are dissolved in the metal *matrix* but do not form a chemical compound with it. Stated more technically, the matrix must remain metallically bonded and must continue to exhibit its characteristic crystal structure. The need to avoid the formation of a chemical compound places rather strict limits on the physical process of alloying, as will be explained more fully later.

Since alloying is essentially a dissolving process in which various elements exist together, the size of the atoms involved is an important consideration. Recalling the analogy of stacking balls in a box, consider what would happen if some were basketballs and some were baseballs. Obviously the size difference precludes the uniformity and regularity of spacing that are requirements of the metallic crystal structure. It has been determined that it is necessary for atoms to be within 8% of

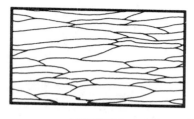
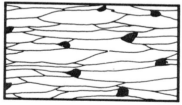
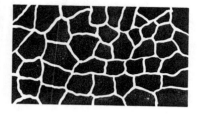

Inco

2-20. Effect of cold-work and annealing on grain size. Cold-working, such as cold-rolling, tends to deform the grains. The resultant grain structure contains grains elongated in the direction of working. If heat is applied to the worked structure, a new grain structure can grow from the worked structure. The new grains are undistorted. This process, called recrystallization, frequently occurs during annealing.

each other in size (atomic diameter) to be mixed or alloyed together in any proportion. If the size variation is over 8% but less than 15%, the atoms can be mixed in limited proportions without destroying the lattice structure. Greater size differences prevent alloying.

Another limitation imposed by crystal structure requirements is that the atoms must form the same type of crystallographic arrangement as in the elemental state: BCC, FCC, or HCP. When both size and crystallographic requirements are satisfied, the alloy that can be made is called a *substitutional solid solution*. Fig. 2-21.

If the two types of atoms differ greatly in size, the smaller atoms can fit into the naturally occurring spaces or "holes" between the larger atoms in the crystal structure. Fig. 2-21. This type of alloy is termed an *interstitial solid solution*. Only elements with very small atoms, such as carbon, oxygen, nitrogen, and hydrogen, can be interstitially dissolved into metals. The gas atoms in the air can become dissolved in metals during processing and act as harmful impurities. To control gas impurities, metals are industrially vacuum-treated. For example, nickel alloys can be melted and heat-treated in a vacuum, which prevents access of gas atoms to the metal matrix.

Intermetallic Compounds

Bonding has been considered in terms of *similar* atoms giving up their electrons to an electron cloud. However, if *unlike* atoms are brought into close contact, there is the possibility that they can chemically interact and bond either ionically or covalently, even if the two atoms are metals. Such chemical compounds are called *intermetallic compounds*. Several common intermetallic compounds are iron carbide (Fe_3C), found in steel or cast iron, and the compound of nickel and aluminum (Ni_3Al), found in many high-temperature nickel alloys.

For two elements to alloy as a metal they must not chemically react, since a chemical compound has substantially different properties than does a metal. A

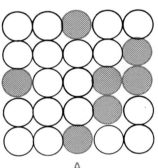

A
SUBSTITUTIONAL

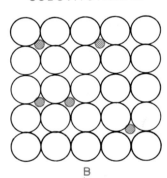

B
INTERSTITIAL

2-21. Solid solutions. A substitutional solid solution occurs when two elements with atoms of nearly the same size are alloyed together. The atoms can assume any lattice site and can be substituted for each other without excessive distortion of the crystal structure. When a large size difference occurs, the small atoms can assume the sites that occur in the voids between the larger atoms. This type of solid solution is termed "interstitial."

Technology of Industrial Materials

chemical compound has a fixed composition by weight, such as 22% nickel and 78% aluminum. It is possible for alloys made of other compositions to exist without the formation of the chemical compound—for example, an alloy of 10% nickel and 90% aluminum. But if an alloy is produced in which nickel and aluminum are present in proportions that satisfy the chemical compound's composition, then the compound will form.

Equilibrium Diagrams

It is clear that alloying is complex and that each alloy system will act differently. Fortunately it has been possible to develop a system to record and illustrate what happens in each alloy system when the alloys are slowly heated and cooled. This was accomplished by using *equilibrium diagrams*. These diagrams act as "road maps" that indicate how a specific alloy will behave. Equilibrium diagrams

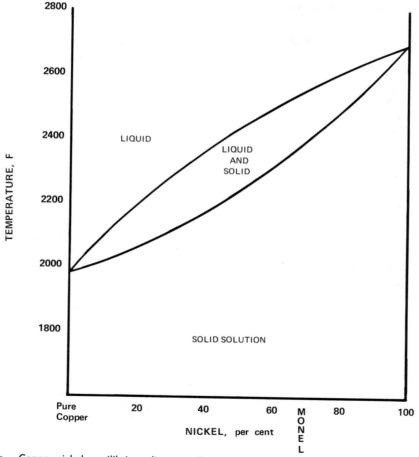

2-22a. Copper-nickel equilibrium diagram. Copper and nickel are completely, substitutionally soluble in each other. Regardless of the relative amounts of copper and nickel atoms present, in its solid phase the alloy will have an FCC crystal structure. Monel alloy, 70% nickel and 30% copper, is a commercial alloy of this system.

Chapter 2. The Nature of Materials

cost, could be used as a ceramic. The graphite allotrope, which exhibits mixed bonding, is commercially used as a refractory—that is, a high temperature-resistant ceramic.

When a ceramic is composed of a chemical compound, the bonding and sizes of the atoms involved produce a more complex structure than is found in metal. The positions and numbers of interacting valence electrons are controlled by the ionic and covalent bonding requirements. However, because the constituent atoms of a ceramic frequently differ considerably in size, complex crystallographic arrangements occur. The complete crystallography of ceramics is beyond the scope of this text, but the common ceramic structures will be examined. (See Chapter 5.)

The ceramic *periclase* (magnesium oxide) exemplifies predominantly ionic bonding. This material is crystalline and has a lattice structure of alternating ions. Fig. 2-25. Other ceramics of this type but with more complex crystal structures are *sphalerite* (ZnS), *corundum* (Al_2O_3), and *fluorite* (CaF_2).

The most widely used ceramics are based on the silicate ion, SiO_4. The primary structural unit of these ceramics is a pyramid or tetrahedral arrangement consisting of the silicon ion occupying the interstitial position among four larger oxygen ions. Fig. 2-26. This unit, called the *silicate tetrahedron*,* is bonded together by a combination of both ionic and covalent bonds, producing great stability. Each tetrahedron is itself an ion with a charge of -4.

Silicate tetrahedra can react in several ways to form ceramics of different types. They can react with metallic ions that donate one electron to each oxygen atom in the tetrahedra. Such a reaction separates the tetrahedra into "islands". The *island silicate structure* is represented by *fosterite*, Mg_2SiO_4. Fig. 2-27.

*Plural: *tetrahedra* or *tetrahedrons*—either is correct.

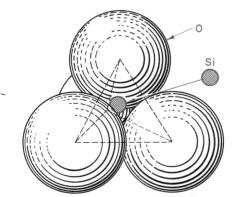

2-26. Silicate tetrahedron. The silicated tetrahedron is formed when a silicon atom is positioned within the void that occurs when four oxygen atoms are stacked to form a pyramid or tetrahedron. This structure has a charge of -4. Reactions between these tetrahedra and other metal ions form many commercial ceramics.

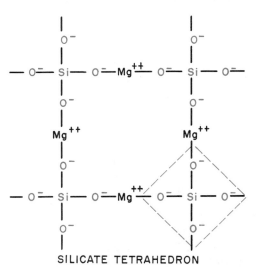

SILICATE TETRAHEDRON

2-27. Island structure of silicates. Fosterite, formed when silicate tetrahedra react with magnesium, typifies the "island structure." The tetrahedra are separated into discrete islands by the magnesium ions.

57

Technology of Industrial Materials

Silicate tetrahedra can also form a *chain structure*. Fig. 2-28. They share oxygen ions to produce the chain. The chains are bonded together ionically by metallic ions, such as sodium or magnesium. It is possible to cleave or break the chains apart at the ionic bond sites because these bonds are somewhat weaker than the covalent bonds. Fibrous ceramics, such as asbestos, have this type of structure. Silicate chains are sometimes called inorganic polymers. (This is because of their resemblance to true polymers, which are discussed in Chapter 4.)

Chains of bonded silicates can extend into two dimensions to form a sheet structure. Ionic bonds at the joint between the two layers are points of lower strength, permitting movement of the sheets. This movement may be cleavage, as in mica, or sliding, as in clay and talc. The lubricating qualities of talc are based on the sliding. Similarly, the ability of a surface coating of clay to be applied to sand grains and then to rebond the clay together is a function of sheet sliding. This is why clay is used as a "binder" for sand molds in foundry. Fig. 2-29.

Finally, the large units of silicate compounds are hard to move. When the temperature of many molten silicates is suddenly dropped below the melting point, the primary silicate units do not have the time to rearrange themselves in the structure of the solid state. At these lower temperatures, molecular motion is too slow to permit this arragement to occur, resulting in the liquid structure being "frozen" in the solid state. This *supercooled liquid,* a glassy material, acts like a solid in many ways, but is in reality a *liquid*. Window glass, composed mainly of silicates, is an example of this structure.

Organics

Materials used industrially that are organic in nature are very numerous. These include plastics, rubbers, woods, and plant or animal fibers. All organic materials are composed of very large covalently bonded molecules generally based on carbon.

The large molecules are composed of a repetitive pattern of structural units called *mers*. Hence, the chemical name *polymers* (many mers). A representative polymer molecule can be considered as a chain, with each mer acting as a link. Fig. 2-30. A mass of these chains can be considered a *polymer aggregate*. The properties of a particular polymer aggregate depend on the manner in which these chains are linked together.

When there is only slight chemical attraction between the chains, they are free to slide over one another if a very slight load is applied. Only the relatively weak van der Waals forces hold the chains together as a mass. Natural rubber—such as a rubber band—and polyethylene are examples of this structure.

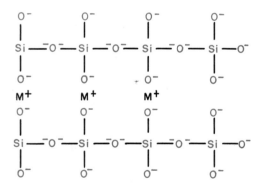

CHAIN OF TETRAHEDRA

2-28. Chain structure of silicate tetrahedra. The silicate tetrahedra share oxygen ions to form a chain. The chains can be ionically bonded together by metal ions, shown as M+. This bond can be broken easily, resulting in a fibrous material such as asbestos. Note similarity to polymer chains. (See Fig. 2-9.)

2-29. Sheet structure of ceramics. Electron micrograph (9500×) of kaolin clay crystals. The sheets are very thin and lightly bonded together. Under slight pressure, the sheets are caused to slip over each other, thus covering the sand grains.

It is possible to link the chains together chemically with covalent bonds. This will impart much greater strength to the structure and make it more resistant to deformation under load. Vulcanized rubber, such as an automobile tire, and epoxy resins represent the chemically linked chain structure.

In living systems, such as the human body or a tree, the various organs and structural parts are built from very large, complex polymers. Generally the structural parts are attached to each other mechanically by wrapping or intertwining and also chemically where one part is bonded to another. Living systems generally represent composites of two or more major types of polymer materials. Wood, for example, contains strong cellulose fibers held together by lignin for flexibil-

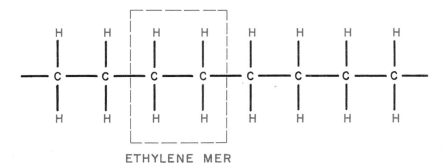

ETHYLENE MER

2-30. Polymer structure. The polymer of the repeating monomer unit *ethylene* is the commercial material *polyethylene*. The monomers are chemically combined to form a large chain molecule.

Technology of Industrial Materials

ity. Human bone tissue is a composite of hard minerals secreted by the body into the strong but soft collagen. The differences among woods can be attributed to differences in specific nature of the polymers of which they primarily consist. The behavior of bone tissue is also largely determined by the nature of its constituent polymers.

BEHAVIOR OF INDUSTRIAL MATERIALS

Behavior under Load

A material that has a crystalline structure, such as a metal or an ionic ceramic, will respond to an applied tensile or compressive load in the manner described on pare 42. It is the lattice structure of these materials which determines this response. As explained, ions in the lattice are held at their equilibrium spacing by the action of interatomic electronic forces. These forces have a springlike effect. Fig. 2-31. At equilibrium, the "springs" are under no load. Applying a tensile load will tend to pull the atoms apart, but the springlike bonding forces have an opposite effect on the atoms, tending to pull them back together. Therefore a sufficiently great force must be applied to obtain a permanent change of shape. Similarly an applied compressive force, tending to squeeze the atoms together, will be opposed by a springlike restoring force. When small loads in the elastic area are applied and then removed, the "springs" cause the atoms to return to their equilibrium spacing, and the piece returns to its original shape. Thus it can be seen that the similar behavior of crystalline materials under both tensile and compressive loads is a result of the structure of these materials. In other words, the elastic properties of crystalline materials are structure dependent.

Metals require an unusually high load or stress to produce an elastic change in interatomic spacing. This property is expressed by the Elastic Modulus (Young's Modulus). For metals, about 10 million to 30 million pounds per square inch of stress are required for each inch of deformation. In another sense, to cause an elastic deformation of only 0.0001" in a piece of steel, a force of about 3000 pounds must be applied to each square inch of cross-sectional area. Therefore metals have an exceedingly high structural integrity of size and shape under load.

Unlike a crystalline material, the glassy materials or the weakly bonded polymers

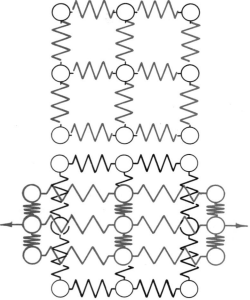

2-31. Elastic response to tensile load. The bonds between metal atoms in a crystal act like springs. The atoms are prevented from changing position by the balance of attractive and repulsive forces within the lattice. A tensile load acts to stretch the "springs" in the load direction and compress them in the opposite direction. When the load is removed, the "springs" go back to their original positions. This response to applied loads continues until the load exceeds the elastic limit.

tend to flow under an applied load. The structural units of these materials can readily move with respect to each other and are not held in position by the same restoring force. As long as the load is applied, the flow will continue and the size of the part will constantly be changing. The rate of change depends greatly on temperature. In the supercooled liquid temperature region, the flow rate is so very low that glassy materials appear to act like metals. However, at higher temperatures these materials exhibit viscous flow similar to a liquid.

Through tensile testing (see Chapter 8) the elastic limit of crystalline materials can be found. This limit is the point beyond which a material under tensile load will yield and permanently deform. Immediately after yielding, brittle metals and most ceramics will break abruptly. However, many ductile metals will elongate up to 60% prior to fracture. Clearly, different mechanisms of deformation must be acting, since the size change in ductile metals is permanent and will not disappear when the load is removed.

Plastic Deformation and Dislocation Theory

The basis for plastic, or permanent, deformation in metals and crystalline materials by the slip mechanism has only recently been explained. The explanation required a new understanding of the lattice, taking account of certain defects in the lattice structure. These defects, termed *dislocations,* occur in all metals. Once these dislocations were understood, it was possible to describe a deformation process in metals wherein the atomic bonds were not broken (since the metal was not broken, merely deformed). The deformation occurred only above a certain stress level (to account for the yield point).

While a full discussion of dislocation theory is beyond the intent of this text, several of its salient features will be described. A dislocation is a missing partial row or an extra partial row of atoms within a crystal. Fig. 2-32. As shown, the absence of some atoms not only leaves an incomplete bond but also affects the spacing of certain other atoms. In other words, some atoms are dislocated from the equilibrium positions where they would have been held if the row were complete.

Although dislocations are fixed in a crystal structure, under an applied load they can move through the lattice to the surface of the crystal, but only in certain directions and only within certain planes called *slip planes.*

When a dislocation reaches the surface, it produces a permanent change in the size of spaces between atoms in the crystal. Fig. 2-33. The deformation process is called *slip* because it appears that one part of the crystal has "slipped" or moved over the remainder of the crystal.

Note that after the dislocation has reached the surface, the crystal contains

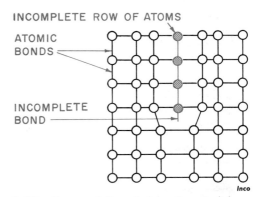

2-32. Representation of dislocation. A dislocation can be represented as either an extra partial row or a missing partial row of atoms in a crystal. The dislocation occurs where the incomplete atomic bond is found and where the atoms consequently are not in equilibrium lattice sites.

Technology of Industrial Materials

only an undeformed and undistorted structure. This would seem to indicate that plastic deformation would stop when all the dislocations have reached the surface of the crystal. However, since each dislocation produces a size change of only one atomic spacing, the large amount of plastic deformation that is observed in real materials must be caused by the generation of new dislocations as the original dislocation moves to the surface of the crystal. The generation of large numbers of new dislocations and their subsequent movement to the surface can only occur if the applied stress is constantly increased. The increased stress is required literally to drive the dislocations through the structure and through other newly generated dislocations that may attempt to block each other. During plastic deformation, if the load is held constant, the deformation ceases until a higher load is applied to drive more dislocations through blocks and to generate more dislocations, thus continuing the process. As the tensile testing of ductile materials proceeds, the reduction of the area under stress is so great that eventually there is insufficient material to carry the load and the material fails.

Strengthening Metals

The yield point indicates the highest stress a material will bear and still retain its original shape. Metals are strengthened by increasing the yield stress. Since the

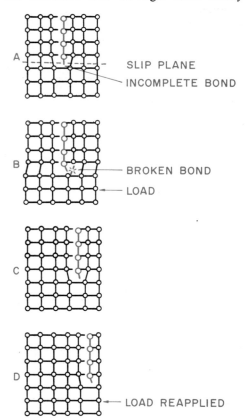

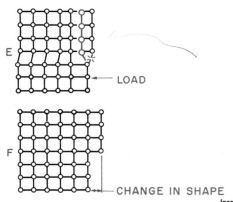

2-33. Dislocation mechanism of plastic deformation. (A) With no applied load, an edge dislocation is present within a crystal but cannot move along the slip plane to the surface; thus no deformation is apparent. (B) Under an applied load, a dislocation can move on a slip plane. (C) If the load is removed, dislocation motion ceases. Note that there is still no apparent plastic deformation. (D) The load is reapplied, causing further dislocation movement. (E) Dislocation reaching surface of crystal. (F) Permanent, plastic deformation has occurred in the size and shape of the crystal when the dislocation has reached the surface of the crystal. Applying an equal load in the opposite direction will not restore the crystal to its original size and shape.

Chapter 2. The Nature of Materials

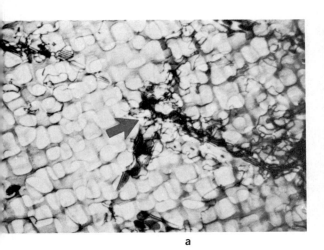

a

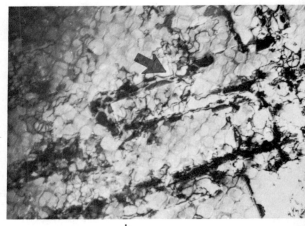

b

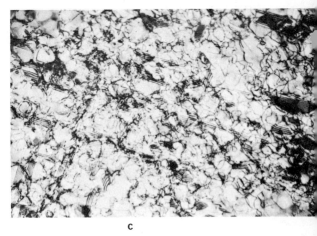

c

2-34. Cold-working. In the upper view (a) the metal has been plastically deformed 0.35%. Slip has occurred as the dislocations moved along slip planes. The dislocations intersected (arrow) and required a higher applied load for further deformation. In view (b) a higher load has been applied and the specimen has been deformed 0.80%. The higher load produced more dislocations, many of which reached the surface. Then deformation ceased because of the intersection (arrow) and will not continue until a yet higher load is applied. After further loading which has caused 1% deformation (c), the metal contains many dislocation intersections.

yield point also marks the onset of dislocation motion, strengthening consists of preventing dislocation motion. Strengthening is accomplished by producing "blocks" within the crystal structure which inhibit dislocation motion. Four industrial methods of blocking dislocation motion are illustrated. Figs. 2-34 to 2-37.

Cold-working—plastic deformation at room temperature—produces a large number of dislocations in the metal structure. The dislocations block each other as they attempt to move along the available slip planes. To drive dislocations through the blocks, higher stresses must be applied. A tensile test of common steel illustrates the effect of cold-working on strength. If the test were stopped at stress of 50,000 psi—which is above the average yield point of about 35,000 psi—deformation would also cease. But the steel structure would retain all the dislocations produced by the initial plastic deformation. However, if the same bar were then given to another tester who was unaware of the first test, he would find that deformation would not occur until a load greater than 50,000 psi was applied. Therefore the initial cold-working opera-

Technology of Industrial Materials

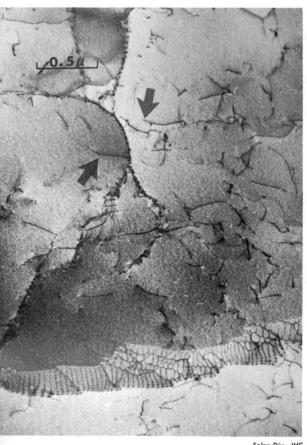

Solar Div., IHC

2-35. Grain boundaries. Dislocations have moved along slip planes to grain boundaries. Unable to cross, they pile up at the boundaries (arrows). Plastic deformation will not occur until a higher stress is applied so that the dislocations can move across the grain boundaries.

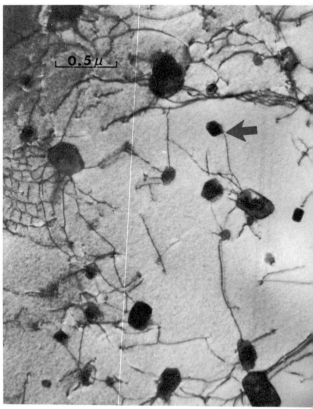

Solar Div., IHC

2-36. Second-phase particles. Dislocations moving through a metal structure under an applied load will pile up when they come to a second-phase particle. (Many examples are shown here, with the arrow indicating a particularly clear one.) Movement ceases until a sufficiently greater load is applied. Only then will the dislocation move through the particle, with resultant deformation.

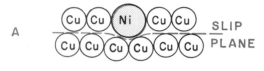

2-37. Alloying. Alloying inhibits dislocation movement by producing local distortions in the slip planes. Such distortions are caused by size differences among atoms of various elements. The addition of a slightly larger substitutional alloy agent (color portion of Fig. A) to a pure metal distorts the lattice. In Fig. B a substantially smaller atom is interstitially alloyed into a pure metal, causing local lattice distortion. Such distortions in slip planes retard dislocation movement and thus increase the yield strength.

tion would have effectively raised the yield strength of the steel and made it capable of withstanding greater stresses without deforming due to dislocation motions. This process is shown in Fig. 2-34.

Because it can be combined with a forming process, cold working is frequently done in industry to raise the strength of metals. Cold-rolled sheet steel and cold-drawn stainless steel tubes are examples of this combined forming and strengthening operation. However, cold-working can also be troublesome. As the yield strength increases, higher loads are needed to carry out each subsequent forming step. Another problem, called *work hardening*, occurs when metals become harder because of the strengthening involved in cold-working.

Grain-size control can be used to strengthen metals. Fine-grained metals are usually preferred for industrial applications because they are stronger than coarse-grained structures. Grain boundaries act as blocks to dislocation motion, preventing them from crossing from one grain to the next. A fine-grained structure contains more boundaries than a coarse-grained structure. Therefore a higher stress must be applied to produce yielding in a fine-grained structure. Many industrial specifications contain limits on grain size. Boundaries are shown in Fig. 2-35.

Producing second-phase particles within the matrix crystal structure is another way of strengthening metals. Fig. 2-36. Small particles, distributed throughout the structure, block the motion of dislocations. To drive dislocations through the second phase particles and thus produce deformation, higher stresses must be applied than the original yield stress of the matrix structure. Second-phase particles can be produced by heat-treating an alloy, such as the precipitation or age hardening of aluminum alloys, or by intentionally adding particles, called *dispersion hardening*.

Alloying increases the strength by distorting the regularity of the crystal structure. Because the solute atoms disrupt the matrix lattice structure, dislocations encounter more resistance to moving in the available slip planes. A higher stress is required to start dislocations moving to the surface; thus a higher yield strength results. Both substitutional and interstitial alloying strengthens metals. Fig. 2-37.

Softening Metals

As just explained, several forming problems can arise due to excessive hardening or strengthening of metals. These difficulties can be severe, both economically and technically, during industrial processing. Softening heat treatments are used to improve forming characteristics by reducing the yield strength, lowering the hardness, and reducing residual stresses. Annealing, tempering, and stress relieving are industrial terms for softening heat treatments. These treatments reduce the number of dislocations and the lattice distortion.

When a cold-worked lattice structure, distorted and containing many dislocations, is heated, the increased thermal vibrations cause the atoms to move toward equilibrium positions. As the lattice distortion diminishes, the residual stress level of the piece decreases. If the heat-treating temperature is made sufficiently high, recrystallization can produce a stress-free lattice and remove the strengthening and hardening effects of cold-working. To soften steel, the metal is heated into the austenite (FCC) region where the new, undistorted lattice forms, and is then slowly cooled.

Blocks to dislocation motion can be removed or reduced by heat-treating. Sec-

Technology of Industrial Materials

ond-phase particles can be dissolved into the matrix by *solution heat-treating*. During annealing, the grain size usually increases, resulting in lower yield strength. Substitutional and interstitial alloys are heat-treated to permit the solute atoms to find minimally distorting positions within the matrix material.

Softening plays another part in making metals more reliable in service. When the residual stress and dislocation pile-ups are reduced, the material will become tougher. Softened metals are more resistant to fracture, as the following discussion will clarify. Thus metals are usually softened to some extent after strengthening to improve their performance in service.

Fracture

Reading about or observing mechanical tests may make it appear that deformation and fracture are closely related. But as materials scientists examined the mechanism of fracturing, they learned that fracture or failure is a process that competes with deformation. Fractures usually originate from pre-existing defects in the material, such as cracks or foreign inclusions. Fig. 2-38. These defects often occur in the commercial production of the material. Under a specific applied load, the pre-existing defects can propagate (spread) as cracks, resulting in failure. The amount of applied load required to start crack propagation depends on the size of the pre-existing defect. When the applied load exceeds the yield point and produces many moving dislocations, some dislocations can pile up at a defect, adding to the defect size and leading to fracturing at lower stresses. The apparent size of the defect increases and the defect can spread, leading to failure at this lower stress. Large dislocation pile-ups can also act as a defect and initiate a crack.

Whether material will deform rather than fracture depends on dislocation movement. As long as dislocations can be prevented from piling up to form defects, the material will deform. It is only when dislocation motion is impeded that fracture results. To prevent premature fracture, industrial components are nondestructively inspected to detect unacceptably large pre-existing defects.

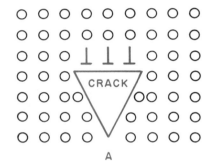
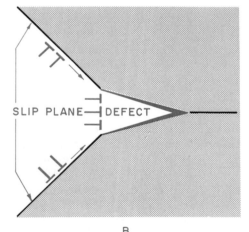

2-38. Fracture. Dislocation pile-ups can result in fracture by producing a crack-like defect sufficiently large to propagate (spread). A pre-existing crack (shown in color) blocks dislocation motion (A). The pile-up (shown in color) adds to the size of the defect. The resulting crack may be large enough to spread under the applied load. In (B) a pile-up (shown in color) itself acts as a defect which can initiate failure under the existing load.

Hot-working is an industrial example of the competitive nature of failure and deformation. Hot-working is accomplished by heating the material to about half its melting point and then deforming it at this temperature. At hot-working temperatures, dislocations can move very easily because more planes serve as active slip planes. Therefore, if dislocations are blocked in one slip plane, they can readily find an intersecting slip plane which has become active and continue their motion to the surface. By reducing the opportunity to form pile-ups at defects, the material can be altered so it tends to deform, not fracture. Hot-working techniques, such as forging and hot-rolling, will produce greater deformation with less chance of cracking than cold-working. However, the material will not be strengthened by hot work.

Certain metals and most ceramics are brittle, fracturing rather than deforming just above the yield point. Such metals may have been processed to increase the yield stress by preventing dislocation motion, or they may contain a second phase—such as iron carbide in cast iron—that acts as a pre-existing defect. Ceramics generally have a somewhat irregular crystal structure because of the size difference of the constituent atoms. Also they exhibit poor fit between grains and they resist dislocation movement at room temperature. For these reasons the lack of ductility under load is not unexpected.

Behavior under Applied Field

Just as the structure of a material determines the response of that material to an applied mechanical load or stress, so does structure affect the response of a material to an applied electrical field. Electrical conduction and magnetic qualities are of primary importance in electrical technology.

Electrical Conduction

The flow of charged particles through a material under the influence of an applied external field results in what is commonly called the flow of current. The relative ease or difficulty with which this flow occurs in a specific material is defined as the *electrical conductivity* of that material. If the flow of charged particles through a material is very restricted, it is easier to measure the *resistance* of that material. However, resistance and conductivity are merely the reciprocal of each other, expressed by the formula:

$$\frac{1}{\text{resistance}} = \text{conductivity}$$

In a solid, the charged particles available for conduction are the ions and the valence electrons. Ions are very bulky and are generally fixed in position by the attractive forces of nearby ions. Therefore the amount of conduction that ions can perform is extremely limited in a solid. Electrons, however, are small and can travel at high velocities, so they can serve as very efficient conductors. The ability of electrons to enter into electrical conduction in various types of solids will be examined.

Electrons in an ionically bonded material are very tightly fixed in their orbits because all the electron shells are completed. In such a position of stability and low energy it is very unlikely that any electrons from the ions will be available to carry the current. A similar situation exists in the covalent bond, in which the electrons are fixed in orbits that are shared by the bonded atoms. In a metal, however, the valence electrons in the electron cloud are able to move quite freely from atom

Technology of Industrial Materials

to atom. This is the cause of the electrical conductivity of metals.

When a metallic conductor is placed in an electrical circuit, the current flow takes place as shown in Fig. 2-39. The circuit consists of a battery and copper wire. The battery serves two purposes: First, it establishes an electrical charge difference in the circuit, with one end becoming negative and the other positive. The effect of this charge differential is to supply a driving force that will cause the electrons to move in a specific direction. The magnitude of the driving force is measured by the potential or voltage of the battery. Second, the battery supplies electrons to the conductor. If you throw the switch, this is what happens:

The positive charge of the battery attracts the negatively charged electrons in the electron cloud. This attraction supplies directionality to the previously random motion of the electrons. On the other end of the wire, the negative charge of the battery repels the electrons, thus reinforcing the directionality of electron motion. Equally important is the battery's ability to supply electrons into the wire from the negative terminal.

The sum of these effects is to produce a condition that permits electrons to flow out of the electron cloud into the positive terminal of the battery. The electrical neutrality of the metal structure is maintained by electrons flowing into the metal from the negative terminal at the same rate at which they flow out. The flow of charged particles is observed as the flow of current.

Temperature exerts an effect on the conductivity of metals. As the temperature increases, the *vibrational amplitude* of the atoms (or ions) in the metal also increases. (Essentially, this means that the ions move more actively.) To help you picture this situation, it could be said that the increased amplitude of vibration serves to decrease the size of the "pathways" through which the electrons can flow. If an electron, while passing through the lattice, happens to "hit" an ion, the electron will rebound. Since such electrons rebound in random directions, they do not usually continue in the direction in which they had been going, and thus they do not contribute to the flow of the current. Understandably, such collisions take place more frequently as the ions move more actively. This explains the fact that the electrical resistance of metals increases with a rise in temperature.

Actually any operation that distorts the structure will tend to increase the chance of electrons "hitting" ions; thus a plastically deformed metal will have a lower conductivity than a non-deformed, equilibrium structure of the same material. Similarly any influence on the metal that produces a flow of electrons will produce a current. The thermoelectric effect of heat

2-39. Electrical conduction in a metal. The metallic bond permits the passage of electrons (shown in color) in response to the attraction of the positive pole of the battery. To maintain electronic equilibrium, the battery must supply electrons to the metal (as shown in color). The flow of electrons is the current in the wire.

on metals is an example of this phenomenon. By heating one end of a metal wire, the electrons in that end of the wire obtain more energy than the "cold" electrons. At higher energies, the electrons can move more readily, and so they move to the colder portion of the wire. This transfer of heat energy is measured as the *heat conductivity* of the metal. Since this type of conduction involves the movement of electrons in a specific direction, the electric current flows through the wires. The size of the current is related to the difference in temperature of the two ends of the wire. The thermocouple makes use of this principle to measure temperature.

Several metals, notably tellurium and selenium, at times exhibit poor electrical conductivity. It appears that these metals contain a substantial amount of covalent bonding within certain temperature ranges; thus they do not always have the prerequisite metallic structure for conductivity.

In liquids, the flow of electric current is carried by ions. These may be ions dissolved within the liquid or they may be the ions of the liquid itself. Sulfuric acid will ionize into hydrogen (H^+) ions and sulfate (SO_4^{-2}) ions as a liquid. These ions, when affected by an applied charge, will conduct current. Water ionizes to an extremely slight extent; therefore pure water is not a conducting material. However, if an ionizable substance such as salt (Na^+, Cl^-) is dissolved in water, the resulting ions will conduct current. Most tapwater contains a sufficient number of dissolved mineral salts from the earth to act as a weak conductor.

Magnetism

Man has known about magnetic phenomena such as the lodestone and the compass for centuries. Only within the last century, however, have scientists been able to explain many of the complexities involved in magnetism.

The explanation of magnetism is closely allied with atomic theory. As explained earlier, electrons are moving electrical charges. They have both spinning and orbital motions. The movement of electrons (1) causes a magnetic field to exist and (2) gives an orientation, either clockwise or anticlockwise, to the field. Thus atoms, since they contain electrons, have a magnetic field. In all materials the directions of the magnetic fields tend to be randomly oriented until an external magnetic field is applied. Then the fields generated by the electrons and atoms within the material may tend to align with the applied field. This causes the material to exhibit magnetic properties. Materials differ in their capacity for such alignment, which accounts for the differing magnetic behavior of various materials.

Strong Magnetism

Strong magnetic behavior, called *ferromagnetism*, is limited to three elements—the metals—iron, cobalt, and nickel. These are the metals which are most important industrially as far as magnetic effects are concerned. Some metallic oxides can also exhibit a form of magnetism, called *ferrimagnetism*, still considered strong but somewhat weaker than *ferromagnetism*.

Atoms of ferromagnetic metals have several available subshells. Because of this, the spins of valence electrons will readily become aligned. The reason for this is that the valence electrons do not have to fill the same shell by adopting antiparallel spins. In iron atoms, for example, the valence electrons can assume positions in either the 3s or 4d subshells because the energy levels of these two subshells are about equal. Thus when an

Technology of Industrial Materials

external field is applied the valence electrons take positions in which their spin directions can and do align with the applied field. This is why iron is strongly magnetic.

When alignment of electron spin direction takes place, *domains* are formed. Domains are regions within the structure of a material where all the constituent atoms

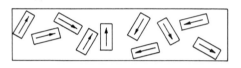

UNMAGNETIZED

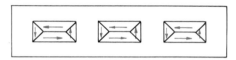

APPLIED MAGNETIZING FIELD

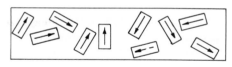

SOFT

HARD OR PERMANENT

2-40. Ferromagnetism. The domains of a ferromagnetic material, in the unmagnetized state, are randomly oriented and cancel one another. When a magnetizing field is applied, the domains reorient themselves into closed magnetic loops which are directionally cooperative. When the external field is removed, a magnetically soft material loses its domain alignment and its magnetism. A hard or permanent magnet retains the cooperative domain structure and thus remains magnetized.

cooperate magnetically. Neighboring domains reorient themselves by aligning in sets to form closed loops. Thus the entire structure cooperates to exhibit a strong magnetic force. This is ferromagnetism. Fig. 2-40.

Once the domains of some ferromagnetic materials are aligned, they tend to retain the alignment after the applied field is removed. These materials are *hard* or *permanent magnets.* Domains of permanent ferromagnets can be destroyed by heating above a specific temperature. This temperature, which varies from one material to another, is called the *Curie Point.* For example, above about 1400° F, iron loses its ferromagnetic properties, because the thermally excited atoms vibrate excessively and disrupt the domain structure.

The domains of other ferromagnetic materials lose their alignments, and thus cancel each other when the applied field is removed. Such materials are *soft magnets.* They are useful in electromagnetic applications when it is desirable for the magnetic force to follow closely the behavior of the magnetizing field.

Weak Magnetism

Most materials are scarcely affected by an applied magnetic field. However, even weak magnetic effects can be important in evaluating materials for industrial use. There are two types of weak magnetic response—*paramagnetism* and *diamagnetism.*

Paramagnetic materials, such as oxygen and silver, exhibit a weak magnetic field of their own if subjected to a strong, externally applied field. In other words, such materials are slightly attracted to a strong magnetic field.

Diamagnetic materials exhibit a weak negative magnetic response—that is they are slightly repulsed away from a strong magnetic field. Diamagnetic materials include copper and zinc.

Effects of Applied Energy

The following paragraphs will discuss the general responses of materials to applied loads. However, the applied loads will be broadly considered as energy. The three most important forms of energy are thermal (heat), mechanical, and electromagnetic.

Earlier passages in this chapter dealt with changes which occur at the atomic or molecular levels when loads are applied—for example, the springlike response of ionically bonded materials to a tensile load. In the following paragraphs attention will be given to changes which occur at the electron level when energy is applied.

Thermal Energy

When the thermal energy level of a material changes through heating or cooling, a wide range of structural responses can occur. For example, a solid material can receive thermal energy by being heated. The atoms—and free electrons, if it is a metal—respond by increasing the period and frequency of their vibrations. This effectively increases the spacing of the atoms and the material expands. But, as Fig. 2-13 indicates, at greater spacing there is less attraction between the atoms. The decreased bonding permits the atoms to move more readily, and thus the changes in structure required for phase or state transformations can occur. When the atoms are tied up in large molecular units, such as polymers, it is the molecules that tend to move. However, their mass makes movement very difficult, so that they do not conduct heat well nor do they vaporize readily.

As heating produces more atomic motion, the atoms or molecules tend to return to equilibrium positions or random orientations. We have seen this effect in terms of stress relief, annealing, and thermal destruction of ferromagnetic domains. Another form of randomizing occurs by a process called *diffusion,* if a difference in chemical composition exists. The carburizing of steel is an example of diffusion. Low-carbon steel is heated in a high-carbon environment, as provided by charcoal. Some carbon atoms from the charcoal will diffuse into the steel structure in an attempt to equalize the differing carbon contents of the steel and the environment. Diffusion processes, such as carburizing, are performed at high temperatures. Such temperatures cause rapid atomic motion, and thus the time required for the process is minimized.

The electronic structure also responds to changes in thermal energy level. It is the thermal response of the electron cloud of a metal that accounts for the high thermal conductivity of that class of materials. Electrons move more rapidly than the much heavier atoms and thus will transfer heat more readily throughout the structure.

Electrons in any material can absorb thermal energy by moving into a higher energy level. But thermally excited electrons are unstable in their new orbit and tend to "drop back" to their original energy level. The energy difference is given off as *electromagnetic radiation.* Color changes which occur when a material is heated are produced by radiation from excited electrons changing energy levels.

Thermionic emission occurs when electrons are sufficiently excited to leave the structure. The highest energy electrons—the valence electrons or the electron cloud of a metal—are the easiest to excite for emission, as in the filament of an electron tube.

Mechanical Energy

Earlier in the chapter, mechanical energy was examined in terms of deformation and fracture. These reactions are at the atomic or molecular level. Another response at the same level is the production of sound waves within a solid. The mechanical energy of the vibratory sound wave causes the structural units of the material to vibrate and thus transmit the sound. Gases, which have no regular structure, and large organic molecules, which are difficult to set in motion, do not transmit sound readily. These materials are useful as acoustical insulators.

Certain complex metallic oxide ceramics have magnetic characteristics that permit applied mechanical energy to produce electrical energy by the *piezoelectric effect.* The crystallographic orientation of the charged structural units comprising these ceramic crystals determines their magnetic effects. An applied mechanical load, pressure, or acoustical vibrations will reorient these units and produce a small electric voltage. An example of the piezoelectric effect is the *transducer,* a device used to measure pressure changes.

Electromagnetic Radiation

Electromagnetic radiation, such as visible light, can promote or excite electrons into higher energy levels. Such radiation supplies energy to the material in specific quantities related to wavelength of the radiation. Electrons can move into a higher energy level only if the quantity of energy supplied is at least as great as the gap between the initial and higher levels. If the incident radiation (that which strikes the surface) does not possess this minimum value by having the proper wavelength, no electronic reaction can occur.

Incident radiation can excite valence electrons sufficiently to leave the material. An example of this is *photoemission* (giving off electrons) which is found in several metals. It is more common, however, for electrons to be promoted to higher energy states within the material. Later, as the excited electrons return to their more stable state, they emit the excess energy as radiation with a wavelength related to the energy difference between their old and new orbits. If the absorption and reradiation of electromagnetic energy is very rapid, it is termed *fluorescence.* However, the electrons can dwell in the excited state for some time before reradiation occurs and they move back to a lower energy level. When this happens the reradiated energy occurs as an "afterglow" termed *phosphorescence.*

Electromagnetic energy can also excite valence electrons into higher energy levels. In some covalent compounds, such as cadmium sulphide, the excitement of the valence electrons results in the ability of the material to carry an electric current. This ability is called *photoconductivity.* The "electric eye" is based on this phenomenon. However, valence-electron excitation can impair covalent bonds. This accounts for the fact that many polymers are chemically degraded by prolonged exposure to light.

Property Sensitivity

Many of the properties we have been discussing are strongly affected not only by the atomic or molecular arrangement

Chapter 2. The Nature of Materials

of a material but also by its gross structure. For example, *grain size, crystalline orientation,* and *direction of prior working* are three common structural variations that greatly affect the actual performance of a material in service. An example is the difference in cutting wood with or across the grain. When a particular material consists of more than one phase or substance, the relative proportion of the constituents and their exact distribution will determine the properties exhibited. For example, the properties of wood are affected by the relative amounts of lignin and cellulose present in the wood. Similarly, steel will vary in mechanical properties based on the relative amounts of soft ferrite and the somewhat harder pearlite eutectoid present in the structure.

The variation of properties across a material is called *anisotropy* and is of major industrial concern. A forging, for example, will exhibit about a 15% higher tensile strength when tested parallel to the direction of metal flow (grain orientation) as compared to tests conducted perpendicular to the direction of metal flow. Magnetization occurs much more readily in certain crystallographic directions. Frequently, magnetic materials are rolled to orient the grains in these directions.

Because of the great effect that the processing and structural variations exert on industrial materials, and because of the variations in performance within a class of materials, each commercially important type of material will be considered in greater detail in the succeeding chapters.

READING LIST

Many of the concepts presented in this chapter were drawn from the following sources. The readings have been classified according to topics of special interest. A number in parentheses after each reference indicates the complexity of the work—the higher the number, the more complex the work in relation to others in the same category.

Materials Science

Guy, A. *Introduction to Materials Science,* New York: McGraw-Hill Book Company, 1972, (5).

Patton, W. J. *Materials in Industry,* Englewood Cliffs, NJ: Prentice-Hall, Inc., 1968, (2).

Ruoff, R. *Introduction to Materials Science,* Englewood Cliffs, NJ: Prentice-Hall, Inc., 1972, (4).

Scientific America. *Materials,* San Francisco: W. H. Freeman and Company, 1967, (1).

Smith, C. *The Science of Engineering Materials,* Englewood Cliffs, NJ: Prentice-Hall, Inc., 1969, (3).

Properties of Materials

Materials Engineering Magazine, *Materials Selector,* annual guide to industrial materials, New York: Reinhold Publishing Corporation, (3).

Parker, E. P. *Materials Data Book,* New York: McGraw-Hill Book Company, 1967, (1).

———. *Modern Materials,* New York: Academic Press Inc., (2). A series of books on developments in industrial materials published in cooperation with the American Society for Testing and Materials, Philadelphia, PA.

Technology of Industrial Materials

Phase Diagrams

American Society for Metals. *Metals Handbook,* Metals Park, OH: ASM, 1948, (2).

Gordon, P. *Principles of Phase Diagrams in Materials Systems,* New York: McGraw-Hill Book Company, 1968, (1).

Alloys

Cottrell, A. H. *Theoretical Structural Metallurgy,* New York: St. Martin's Press, Inc., 1962.

Dislocations

Dieter, G. *Mechanical Metallurgy,* New York: McGraw-Hill Book Company, 1965, (1).

Hirth, J. P. and J. Lothe. *Theory of Dislocations,* New York: McGraw-Hill Book Company, 1968, (2).

Chapter 3

Metallic Materials

Metallic materials are the most important group of industrial materials. Approximately eighty of the 104 known chemical elements are metallic, and at least 50 of them are used industrially to some extent. However, only about 20 metallic materials are found in sufficient quantities to be extracted economically and used extensively. Among those used extensively are iron, aluminum, copper, magnesium, nickel, zinc, lead, tin, chromium, manganese, molybdenum, titanium, tungsten, antimony, gold, silver, and platinum. Even though most of the metallic materials are relatively new, iron, lead, copper, silver, and gold have been used for over a thousand years.

Metallic materials are distinguished from other materials on the basis of their structure. Due to their structure, metals are good conductors of heat and electricity, can be formed plastically without breaking, and can be welded, cast, and machined. Metals generally are strong, heavy, opaque, and lustrous in appearance. Based on chemical composition, metallic materials can be classified into two basic groups: *ferrous* and *nonferrous*. Ferrous metallic materials are those with iron as their principal element. Nonferrous are metallic materials which have little or no iron. The ferrous materials discussed in this chapter are pig and cast irons and steels. Among the nonferrous materials discussed are aluminum, magnesium, copper, zinc, tin, lead, gold and silver, and the refractory metals and alloys.

FERROUS MATERIALS

Ferrous materials are found as iron-bearing minerals deep in the earth or near its crust in many parts of the world. If the iron content of an iron-bearing mineral is sufficient to be commercially mined, it is called *iron ore*. The iron in an iron ore is mixed with such substances as oxygen, sulfur, nitrogen, silicon, phosphorus, and other earthlike impurities. The most common iron ores in the United States are *hematite* with a chemical formula of Fe_2O_3; *magnetite,* a magnetic iron oxide Fe_3O_4; *limonite* Fe_2O_3; and *siderite* $FeCO_3$. Several states have iron ore deposits, but northern Minnesota is the principal domestic source.

Iron ores are extracted from the earth by two mining methods:

◆ *Underground or shaft* mining, used for deep deposits.
◆ *Open-pit* mining, used for deposits near the earth's crust.

In underground mining, vertical shafts as deep as a mile are made near to the iron ore deposit; from the shaft horizontal tunnels (or drifts) extending thousands of feet are opened into the iron ore. The iron ore is blasted loose into the tunnels, hauled to the vertical shaft by underground railway cars, trucks, or conveyors, and lifted to the surface in buckets. Fig. 3-1. In open pit mining the earth's crust is removed with large bulldozers to uncover the iron ore. The uncovered ore is blasted loose, dug up with large power

Technology of Industrial Materials

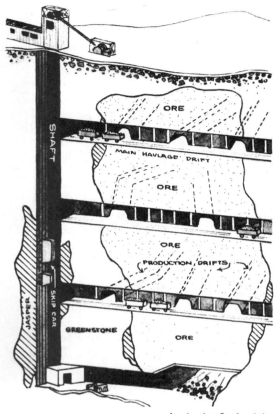

3-1. Schematic representation of a typical underground mine.
American Iron Ore Association

shovels, and hauled out of the mine by trucks, trains, or conveyors. Fig. 3-2.

To increase efficiency and quality and reduce cost of the ironmaking and steelmaking processes, the ore before melting is improved by *benefication* or *concentration* methods. The main aim of these methods is to increase the percentage of iron in the ore by eliminating impurities. Concentration and benefication methods usually include crushing, screening, and washing of the ore for eliminating its impurities. In recent years as the high grade (rich in iron) deposits are gradually depleted, the iron ore industry has developed methods to utilize economically such lower grade iron ores as *taconite* and *jasper* of Minnesota and Michigan. The taconite and jasper iron ores are heavy, hard quartz containing 25 to 40% iron. To make them useful for melting, an expensive concentration process known as *agglomeration* is employed. Four basic steps are involved in the agglomeration process:

1. The ore is extracted and ground to fine powder form.
2. The iron particles are separated from their impurities by magnetic separators, flotation cells, or spiral separators employing the weight or magnetic characteristics of iron.
3. Because the 60 to 65% concentrated iron ore cannot be used in the melting furnace in powder form, it is made into small pellets or balls $\frac{1}{2}$ to $\frac{3}{4}''$ in diameter.
4. The small pellets or balls are heated (sintered) to 2400° F to make them strong for handling purposes. Fig. 3-3.

From the concentration plant or the mine, the ore is transported to the steelmaking plants by railroad cars, barges, or ocean and lake ships to be converted into pig iron by reduction.

Production of Pig Iron

Because all iron ores have a high oxygen content, they are in the form of iron oxide when mined. Therefore the first step in the production of iron is to remove the oxygen by melting the ore in a *blast furnace*. Fig. 3-4. The iron taken from the blast furnace is an impure form called *pig iron*. Pig iron is an intermediate product used to manufacture cast iron and steel. The principal materials used in the blast furnace to produce pig iron are: *iron ore, coke, limestone,* and *hot air*. The iron ore is used to provide the iron; coke to provide heat needed to melt the iron ore and

3-2a. Typical open pit iron ore mine.

to provide carbon monoxide to help remove the oxygen from the iron oxide; limestone to purify the iron by joining with such impurities as sulfur and silicon dioxide; and the hot air to provide the needed oxygen for burning the coke. For a daily production of about 800 tons of pig iron, an average-size blast furnace would require 1,500 tons of iron ore, 600 tons of coke, 250 tons of limestone, 2,800 tons of air, about 50 tons of related materials such as roll scale (the oxide obtained from the rolling mills), and about 6,000,000 gallons of water for cooling purposes.

3-2b. Close-up view of operations in an open pit mine.

Technology of Industrial Materials

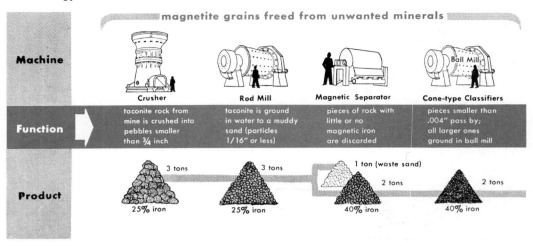

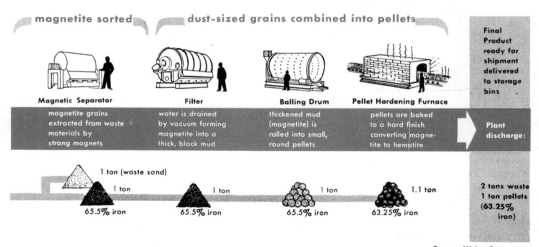

3-3. Flow chart showing the basic steps of taconite preparation for the blast furnace.

The operation of the blast furnace is simple. The iron ore, coke, and limestone in predetermined amounts, forming the *charge* of the furnace, are loaded in carriers called *skip cars* and poured into the furnace from the top. Hot air from the stoves is blown into the furnace near the bottom, causing the coke to burn at about 3000° F. Fig. 3-5. As the iron ore melts it drops to the bottom of the furnace and the charge moves gradually down with new material being added at the top in a continuous process. The hot air moves upward forming the carbon monoxide which helps to remove the oxygen from the iron ore by turning into carbon dioxide. The limestone melts and changes into lime to join with other impurities to make up a substance called *slag* which forms on top of the melted iron. The

Chapter 3. Metallic Materials

Inland Steel Company

3-4. Typical installation of two blast furnaces and their stoves.

melted iron and slag are drawn from the furnace every 4 to 6 hours. The iron is cast into forms called pigs, weighing 100 lbs., or it may be transported in melted form to the steel mill. The slag is used for making cement blocks and for many other applications. The hot gases are collected at the top of the furnace and used in the stoves to preheat the air used in the furnace.

Pig iron contains about 3 or 4% carbon, 0.06 to 0.10% sulfur, 0.10 to 0.50% phosphorus, 1 to 3% silicon, and certain amounts of other impurities. It is soft, brittle, of low tensile strength, possesses

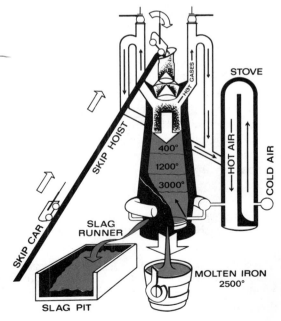

U.S. Steel Corporation

3-5. Schematic representation of a typical blast furnace installation. The tall structures at right and left are the actual furnaces. The silo-like structures in the center are the stoves.

79

high damping capacity (the ability to absorb energy due to vibrations), and looks much like graphite (because of its high carbon content). It is used for making steel and cast irons.

Production of Cast Iron

Cast iron is an alloy of iron, carbon, silicon, and various other elements in insignificant amounts. It is produced by remelting pig iron, scrap cast iron, and steel. Over 20,000,000 tons of cast iron are produced and used annually in various applications in this country. To produce cast iron, a small open-hearth furnace or an air furnace can be used, but most cast iron is produced by the *cupola* furnace. The cupola is similar to the blast furnace. It consists of a vertical stack made with plate steel at the outside and lined with firebricks in the inside. An average cupola would be approximately 5' in diameter and 20' high. The charge of a cupola consists of coke, pig iron, scrap cast iron and steel, limestone, and air. The charge materials are loaded into the furnace from the top and move downward as melting takes place at the bottom. Air is blown in near the bottom of the stack to burn the coke and generate the necessary heat to melt the charge. The cupola furnace does not operate continuously and no significant chemical reactions take place in it. Therefore the main purpose for remelting the pig iron in the cupola is to reduce certain amounts of impurities and increase its uniformity.

The cast iron produced by the cupola furnace contains about 2 to 4% carbon. Because of this relatively high carbon content, and because the carbon is in either the graphite or the combined (carbide) form, this iron is widely used to make castings in a foundry.

The types and many of the properties of cast iron are largely determined by whether the carbon it contains is in the graphite or the carbide form. The form the carbon takes is influenced by the composition of the cast iron, the rate of cooling, and by possible treatment before casting (alloying) and heat treatment after casting. Thus, by controlling these variables, different types of cast iron with unique properties can be produced.

The three primary types of cast iron—*gray, white,* and *nodular* or *ductile*—are produced directly from the cupola furnace. The secondary or *malleable* cast iron is produced by heat-treating the primary cast irons, and the *special alloy* types are produced by alloying the primary cast irons.

Gray Cast Iron

Approximately 70 to 80% of all castings are made with gray cast iron—an alloy containing 2 to 4% carbon, 1 to 3% silicon, and insignificant amounts of manganese and phosphorus. The majority of the carbon in gray cast iron is in the form of free graphite flakes with a very small amount in the combined form as carbide. Therefore the properties of gray cast iron are affected by the size, shape, and distribution of its graphite flakes. Fig. 3-6. Gray cast iron is soft, easy to machine, with excellent damping capacity, but brittle and of low strength. It has high resistance to wear, corrosion, and heat, but low thermal expansion. It is the least expensive of all types of cast irons. Because of this as well as its high damping capacity and ease of casting, it is extensively used for such applications as machine tool bases, frames of machinery, and engine blocks. There are seven ASTM specification gray cast irons—that is, the ASTM (American Society for Testing and Materials) recom-

3-6. Photomicrograph showing distribution and shape of graphite in gray cast iron.

mends seven classes of this metal, based on minimum tensile strength. Table 3-A gives the ASTM classes of gray cast iron and some of their basic properties and typical applications.

White Cast Iron

This metal is produced by the same process as gray cast iron, but its composition is controlled more rigidly to produce a microstructure with the majority of the carbon in the combined form as carbide (cementite). White cast iron gets its name from its whitish metallic appearance when fractured. This is caused by the presence of carbon in its combined form in the microstructure of this metal.

The properties of white cast iron are affected by such factors as amount and form of carbon in its microstructure; amount of silicon, manganese, phosphorus, and various other elements; cooling rate and heat treatment. The carbon content affects hardness—the higher the carbon content, the harder the iron. The silicon content affects the formation of iron carbide. The cooling rate affects wear resistance. Manganese and phosphorus affect the structure and their presence is kept to a minimum. Such alloying elements as chromium, nickel, copper, and molybdenum, singularly or in combination, affect the properties of this metal.

Because the majority of the carbon content is in the combined form, white cast iron is hard (Brinell Hardness No. 350 to 600). It is also brittle and the most difficult to machine of all cast irons. White cast iron is employed primarily to make castings to be transformed into malleable cast iron by heat treatment. Because of its high wear resistance, it is used in such applications as mill liners, crushing equipment, and grinding mills. Even though there are

Technology of Industrial Materials

Table 3-A
Basic properties and typical applications of ASTM specification gray cast irons.

PROPERTIES AND APPLICATIONS									
ASTM CLASS	Tensile Strength, psi*	Bending Load, psi*	Compressive Strength, psi*	Torsional Strength, psi*	Modulus of Elasticity* (Million psi)	Brinell Hardness	Damping Capacity	Machinability	Typical Applications
20	20,000 to 24,000	1,600 to 2,200	80,000 to 85,000	25,000 to 27,000	9.6 to 14.0	160–180	Excellent	Fair	For thin section castings with close dimensions
25	25,000 to 29,000	1,800 to 2,400	95,000 to 100,000	30,000 to 35,000	11.5 to 14.8	170–210	Excellent	Fair	
30	30,000 to 34,000		105,000 to 115,000	40,000 to 45,000	13.0 to 16.4	210–230	Excellent	Fair	For castings of machinery, automotive and water works
35	35,000 to 39,000	2,300 to 3,000	120,000 to 130,000	50,000 to 55,000	15.4 to 17.2	210–230	Excellent	Good	
40	40,000 to 47,000	2,500 to 3,400	135,000 to 150,000	57,000 to 65,000	16.0 to 20.0	210–240	Excellent	Good	For machine tools and heavy motor blocks
50	50,000 to 57,000	3,000 to 4,000	155,000 to 170,000	70,000 to 75,000	18.8 to 22.8	230–270	Good	Excellent	
60	60,000 to 65,000	3,400 to 4,500	175,000 to 190,000	80,000 to 90,000	20.4 to 23.5	250–290	Good	Excellent	For dies, crankshafts, heavy-duty machine tools, presses, etc.

*Values based on 1.2-inch test bar

no standard specifications for white cast irons, some common types are identified as *cupola white cast iron, malleable white cast iron* and various forms of *martensitic nickel-chromium alloy white cast irons.* Although some castings made from white cast iron are used without heat treatment, most of them require such treatment to relieve stresses and improve their properties.

Chilled cast iron is another form of wear-resistant white cast iron. It is produced by pouring molten gray cast iron against a *chill* such as a steel plate in the mold. The surface of the iron in contact with the chill cools rapidly, producing a metal with surface properties similar to those of white cast iron but interior properties like those of gray cast iron.

Nodular Cast Iron

Nodular cast iron (also referred to as ductile or spherulitic graphite) is produced by melting the same raw materials as for

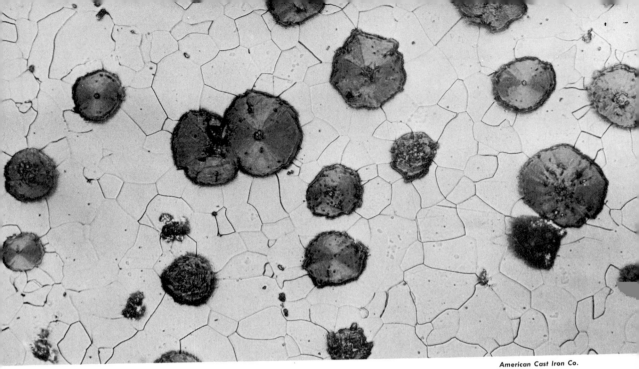

American Cast Iron Co.

3-7 Photomicrograph of nodular cast iron showing carbon content in nodular form.

gray cast iron in a furnace (cupola, air, or electric). Unlike gray and white cast irons, whose carbon content is in the free graphite flake and combined forms, the majority of the carbon content in nodular cast iron is in the nodular graphite form (small ball-shaped carbon formations).

To produce cast iron with carbon content in the nodular graphite form, a two-step process is used:

- First the raw material is melted and a specified amount of magnesium is added to the melt. The addition of magnesium changes the carbon from graphite-flake form into nodular form and some of it into cementite. However, the presence of cementite tends to reduce the ductility of nodular cast iron.
- The second step is to change or eliminate the formation of cementite. This is accomplished by adding or *inoculating* the melt with a small amount (0.5 to 1.5%) of such materials as manganese, cerium, cal-

cium, and sodium. The result after inoculation is a cast iron with the majority of its carbon content in nodular form. (Fig. 3-7).

The composition of nodular cast iron is 2.20 to 4.00% carbon; 1.80 to 2.80% silicon; up to 0.80% manganese; 0.10% maximum phosphorus; and 0.03% maximum sulfur. Nodular cast iron has a ductility of 2 to 15% elongation and a 21 to 25 million psi modulus of elasticity. It is strong, with 50,000 to 100,000 psi tensile strength and 40,000 to 150,000 psi compressive strength. It can be easily machined and has the same wear and corrosion resistance as gray cast iron. Nodular cast iron has a relatively low melting point, good fluidity for casting, and good hardenability and toughness. According to ASTM specification nodular cast irons are designated by a three-number system. The first number indicates the minimum tensile strength (in 1000 psi), the second the yield strength (in 1000 psi), and the last the minimum percent of elongation. Thus a 60-45-10 nodular cast iron is of 60,000 psi minimum tensile strength, 45,000 psi yield strength, and 10% elongation. The three types of specification nodular cast iron mostly used are 60-45-10, 60-40-15 and 60-45-15. Nodular cast iron is extensively used by automotive manufacturers for car and truck parts, by machinery builders for machine bases and frames, by agricultural implement manufacturers, and in similar industrial applications.

Malleable Cast Iron

Malleable iron is considered secondary cast iron because it is produced by heat-treating white cast iron of proper composition. To produce malleable cast iron, first the raw materials are melted in a furnace (cupola, air, or electric) to convert them into white cast iron. This iron is then heat-treated in a two-stage annealing process. Based on the form of the carbon content caused by heat treatment, two basic types of malleable cast iron are produced: the *ferritic* or *standard* and the *pearlitic*. Ferritic malleable cast iron has a microstructure of small nodules of temper carbon in a ferrite matrix. To accomplish this type of microstructure, white cast iron is heated to about 1750° F for 5 to 20 hours to break down the combined carbon. Then it is heated to about 1300° F to convert all carbon content into temper carbon in the ferrite matrix.

To produce pearlitic malleable cast iron, the white cast iron is heated to the same temperature as that for producing ferritic cast iron, cooled slowly to 1600° F., then air-cooled to room temperature. The result is a pearlitic malleable cast iron which has a certain amount of carbon in combined form in the ferrite matrix.

The composition of malleable cast iron is 2.00 to 2.70% carbon; 1.00 to 1.70% silicon; 0.40 to 0.55% manganese; 0.10 to 0.20% sulfur; and 0.03 to 0.10% phosphorus. Malleable cast iron is the most machinable of all cast irons. It is ductile, with 5 to 20% elongation and 23 to 25 million psi modulus of elasticity. It is strong (40,000 to 100,000 psi tensile strength) and hard (Bhn 110 to 145). Malleable cast iron is high in dumping capacity, toughness, and corrosion resistance, and is uniform in its structure because of heat treatment. Malleable cast iron is extensively used by automotive, machinery, agricultural equipment, hand tool, and hardware manufacturers.

Alloy Cast Irons

When the content of such alloying elements as silicon, nickel, chromium, aluminum, and molybdenum is appreciably high in a cast iron, the metal which results is an *alloy cast iron*. The main purpose of alloying cast iron is to change its structure, and hence its properties, to fit specific service conditions. Many types of alloy cast iron are produced with high resistance to corrosion, heat, and wear. The corrosion-resistant alloy cast irons are of the high-silicon type (silicon above 14.0%), high-chromium type (chromium 20 to 30%) and high-nickel type (nickel 14 to 30%). The heat-resistant alloy cast irons are made by adding chromium, nickel, molybdenum, aluminum, and silicon, singularly or in combination and in excess of 3%, to gray cast iron. The wear-resistant cast irons are made by adding alloying elements to white cast iron.

Production of Steel

Steel is the most important of all ferrous materials and possibly of all materials used in manufacturing and construction.

Today steel is produced mainly by four basic processes:

- Open-hearth.
- Basic-oxygen.
- Electric.
- Bessemer.

Although most steel tonnage is produced by the open-hearth process, the basic-oxygen process (BOF or BOP) is rapidly growing in importance. It is estimated that in 20 years or less, basic oxygen will replace the open-hearth process for the production of carbon steel. The electric steelmaking process is also becoming more important as the demand for more specialized types of steel is growing to meet the needs of modern technology. The Bessemer process, once the principal steelmaking method, is largely obsolete; very little steel is now produced by this method.

Regardless of the process used, the production of steel from pig iron and scrap involves three basic functions:

- Removing as much impurity as possible.
- Adjusting carbon content to meet desirable specifications of steel.
- Facilitating the addition of required alloy elements.

The Open-Hearth Steelmaking Process

Originally developed in England in the 1860's, this soon became the principal steelmaking process throughout the world. For almost a century, over 90% of the world's steel has been produced by this process, which employs an open-hearth furnace in which the metal is open to the sweep of the flames across the hearth. Fig. 3-8. The furnace is considered both *reverberatory* and *regenerative*. It is considered reverberatory because the metal in the hearth is partly heated by radiation from the roof. It is considered regenerative because the hot gases produced in the furnace are used to preheat the fuel by passing them through regenerative chambers (checkers) which store part of their heat before they are released to the atmosphere. The open-hearth process is divided into *basic* and *acid* processes. In the United States the basic open-hearth process is the more common.

The open-hearth furnace is built on two levels. The hearth, charging area, and

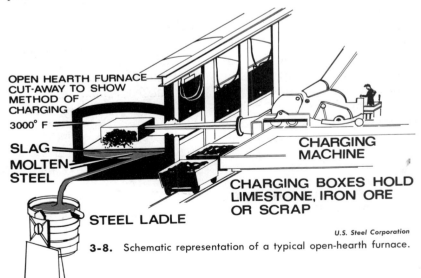

U.S. Steel Corporation

3-8. Schematic representation of a typical open-hearth furnace.

controls are located on the top level, and the rest of the furnace is on the lower level. The hearth—the part of the furnace where melting and refining of the metal takes place—is a shallow, rectangular basin about 90' long and 30' wide. The regenerative chambers or checkers made of special bricks are used to store heat for preheating the fuels and are built on both sides of the furnace. While the checkers of one side are heated from the outgoing hot gases, those on the other side heat the incoming fuel and air; direction of flow is reversed every 30 minutes to increase the economy of furnace operation.

A single furnace load is called a *heat*. Heats vary in size. Although few furnaces can accommodate heats up to 600 tons, most heats are of 100 to 300 tons. The materials used to make steel in the open-hearth furnace are pig iron, scrap metal, limestone, iron ore, fuel (petroleum or gas), oxygen, and air. The limestone is used as a flux because it combines with impurities to form the slag. The scrap metal, which makes up 50% of the metallic charge, is primarily steel scrap and is charged cold in the furnace. The pig iron, the other 50% of the charge, is charged in the furnace cold, molten, or both. The oxygen is used to reduce the carbon content by oxidation and to increase the flame temperature to about 3000° F. Under high pressure it is introduced onto the molten metal in the hearth through a retractable steel pipe known as the *oxygen lance*. The iron ore (or sometimes mill scale) is used to oxidize and refine the melt.

Various practices are used to operate the open-hearth furnace, and they may vary from heat to heat. However, the complete cycle, from charge to tap, for an average-sized heat requires about 8 to 10 hours. Thus, if the heat consists of 500 tons, the production of the furnace would be about 50 tons per hour. After the refining is completed, the charge is tapped through a hole at the rear of the hearth. The melted steel is poured in a ladle and is ready for pouring into ingots. Although the open-hearth process is primarily used for making carbon steel, some types of alloy steel can be made by this method.

The Basic-Oxygen Process

The BOF process is the newest industrial steelmaking method. Even though its widespread use began as recently as the early 1950's, today this method produces about one-third of the world's steel. Its main advantage over the other steelmaking processes is its short time cycle; it requires about 55 minutes to complete a heat. The efficiency of BOF is due to the use of high purity gaseous oxygen as the sole oxidizing agent in refining. The four principal materials used to make steel by the BOF process are molten pig iron (65 to 85% of the charge), steel scrap (15 to 35% of the charge), limestone, and pure (99.5% pure) gaseous oxygen.

The basic-oxygen furnace is relatively simple—a pear-shaped vessel built of steel plate and lined with firebricks. Fig. 3-9. The furnace is supported on trunnions so that it can be tilted about 180°. In charging, the furnace first is tilted and a predetermined amount of scrap is charged. The molten pig iron is charged with a ladle on the top of the scrap and the furnace is raised to a vertical position. A retractable water-cooled oxygen lance is mechanically lowered at a specified distance from the charge and clamped in position. The oxygen is turned on with a pressure of about 150 psi and a flow of 5,500 cubic feet per minute. Coming in contact with the molten metal, the oxygen reacts violently and combines with the carbon in the pig iron to form carbon monoxide which escapes as gases from the top of the furnace.

During the oxidation period, a great amount of heat is generated, raising the temperature of the charge to over 3000° F. While the oxidation goes on, a certain amount of limestone is added through a retractable chute. Under the intense heat the limestone melts, mixes with the oxidized impurities, and comes to the top of the melt as slag. When the oxidation ends, the oxygen is turned off and the lance is raised out of the furnace. The furnace is tilted and samples are taken for chemical analysis. The melted steel, at a temperature of about 2900° F, is discharged by gradually tilting the furnace over a ladle. At the end the slag is removed and the steel in the ladle is ready to be poured into ingots. Although some alloy steel can be produced by the BOF, this process is mostly used to produce carbon steel. As mentioned, the BOF requires only about 55 minutes for a heat. Therefore, if the size of the furnace is 200 tons, the furnace production would be 200 tons per hour, which is substantially more than for a comparable open-hearth furnace. Basic-oxygen furnaces vary in size from 35 to 200 tons.

The Electric Steelmaking Process

While the open-hearth and basic-oxygen steelmaking processes are used primarily to produce most of the common grades of steel, the electric furnace is used to produce special grades such as tool and die steels, stainless, and heat-resistant steels. Unlike the open-hearth and BOF processes, which require fuel and gas to produce heat, the electric furnace employs electric current. Three-phase alternating current is mostly used, and the heat is generated because of the proximity of the electric arc and the resistance of the metallic charge in the furnace. Depending on the type of furnace, the heat may be generated between the three electrodes and the charge or between electrodes. Therefore steel produced by the electric furnace is the cleanest of all steels because no fuel or gases are used that might contaminate the charge.

Although electric furnaces may be of either the *arc* or the *induction* types, far more steel is produced by the arc furnace. An electric arc furnace is a circular shell of steel plate, like a large teacup, lined inside with firebricks. Fig. 3-10. The furnace is mounted on rockers to enable tilting to discharge the molten steel in a ladle. Modern electric arc furnaces are made with a "swing-over" roof which allows for top charging of the furnace. The charging is done with a large bucket with a removable bottom. After the roof has been

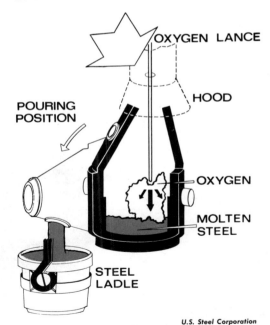

U.S. Steel Corporation

3-9. Schematic representation of the basic oxygen furnace.

Technology of Industrial Materials

swung over, the bucket is loaded with selected steel scrap and suspended over the furnace so the charge can be dropped in from the bottom of the bucket. Through the dome of the roof three large cylindrical electrodes (5 to 25″ in diameter) of carbon or graphite are lowered into the furnace to carry the current to the charge and produce the heat. The distance between the electrodes and the charge, where the electric arc is formed, is automatically kept constant. The sizes of electric arc furnaces may range from 2 to 200 tons. The time required to produce a heat may vary from 3 to 6 hours, depending on the size and type of heat.

In most cases the charge consists of highly selected steel scrap; small amounts of burned lime and some mill scale may be added. If small amounts of alloy elements are needed, they are usually added at the ladle, while large amounts may be added in the furnace. Both *acid* and *basic* methods are used with the electric process, with the acid method primarily for foundry steel and the basic method for special types of steel. The electric furnace has certain advantages over the open-hearth and BOF. It provides high temperatures (up to 3500° F) very rapidly and does not require fuels and gases to operate, thus involving less contamination of the steel. Because close control of the composition is possible in the electric furnace, expensive alloying elements can be added without much loss by oxidation. The main disadvantage of the electric furnace is the high operating cost, due primarily to the cost of electricity used.

The Bessemer steelmaking process has become obsolete since the invention and widespread use of the BOF.

Steel Mill Products

After the steel is tapped from the open-hearth, basic-oxygen, or electric-arc furnace in a ladle, it may be poured into ingot forms; it may be used directly in a continuous casting mill; or it may be used directly in the foundry for steel castings. Before the steel is cast into ingots or used in the continuous casting mill, however, certain *vacuum degassing* and *deoxidizing treatments* are carried out for improving

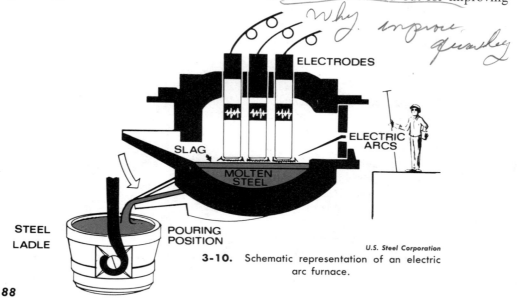

3-10. Schematic representation of an electric arc furnace.

U.S. Steel Corporation

Chapter 3. Metallic Materials

the working characteristics and quality of the metal. Vacuum degassing is done primarily for reducing the hydrogen content of steel, thus preventing the formation of internal cracks (flakes) in certain steel mill products. However, the oxygen and nitrogen contents are also reduced during vacuum degassing. Therefore among the important benefits derived by vacuum degassing are:

◆ Elimination of internal cracks by reducing the hydrogen content.

◆ Increased internal cleanliness by reducing oxide formation due to oxygen content.

◆ Improved steel quality by reducing nitrogen content.

Vacuum degassing can be accomplished by subjecting the molten steel to a high vacuum source as illustrated in Fig. 3-11.

As the steel comes out of the furnace, the chemical reaction between carbon and oxygen continues, causing the gases cre-

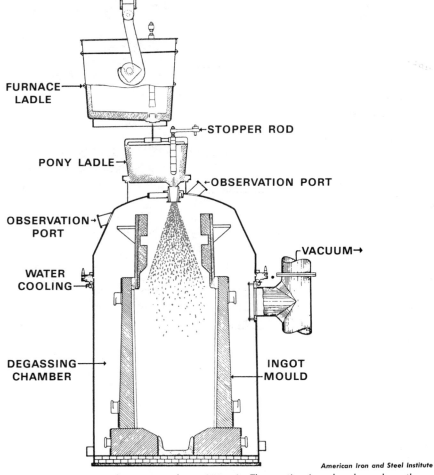

American Iron and Steel Institute

3-11. Schematic representation of a vacuum degassing unit. The portion in color shows how the melted steel is poured into the unit.

89

Technology of Industrial Materials

ated by the chemical reaction to be entrapped in the ingot during solidification. However, as the metal solidifies in the ingot, its internal characteristics vary, resulting in what is known as *segregation*—that is, concentration of certain steel ingredients at certain parts of the ingot. In addition to segregation, a nonuniform solidification known as *pipe* is present in the ingot. Both pipe and segregation affect steel quality and its working characteristics but can be reduced by controlling the gases evolved. Control of the gases is accomplished by adding such deoxidizers as ferrosilicon and aluminum before solidification starts.

Based on the extent of deoxidation done before solidification, four types of steel are usually produced:

- *Rimmed steel* is produced by slightly deoxidizing the melt upon pouring it into the ingot. This type of steel is better suited for applications requiring high surface quality and cold forming.
- *Killed steel* is produced by deoxidizing the melt completely, thus stopping the bubbling in the melt. This results in high uniformity of composition and in properties desirable for such applications as forging, piercing, and heat treatment.
- *Semikilled steel* is produced by an intermediate deoxidation between rimmed and killed steels.
- *Capped steel* is produced by the same deoxidizing process as rimmed steel, but the ingot is covered (or capped) after pouring to reduce the rimming action in the ingot.

Rolling of Metallic Materials

Though some materials are available in the ingot form, most metallic materials are made available to users in the form of finished mill products such as sheet and strip, bar and plates, pipe and tube, struc-

Inland Steel Company
3-12. Stripping steel ingots from their molds.

tural shapes, wire, and rails, to mention only a few. *Rolling* is the basic process used to transform a metallic material from ingot into a finished mill product. Rolling consists of passing the metallic material between two rolls which revolve at the same speed but in opposite directions. Because the opening between the rolls is smaller than the thickness of the material passing through, rolling shapes the material, reduces its cross-sectional area, and increases its length. If the material is rolled at room temperature, the process is called *cold-rolling* and the product is referred to as *cold-rolled steel;* if the material is rolled hot, the process is called *hot-rolling* and the product is *hot-rolled steel.* Cold-rolling is primarily a finishing process while hot-rolling is a shaping process.

The steel ingot, stripped from its mold, Fig. 3-12, is placed in a furnace known as the *soaking pit* and is heated to about 2200° F. This temperature makes the metal plastic so it can be shaped by rolling. Fig. 3-13 illustrates the basic steps

Chapter 3. Metallic Materials

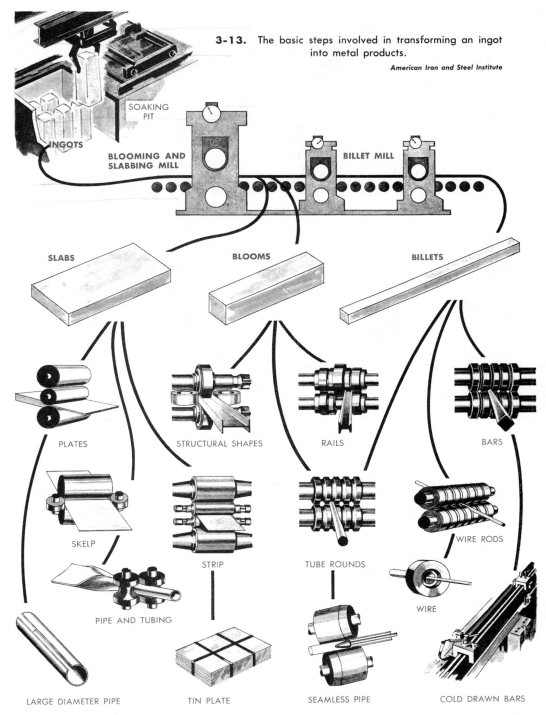

3-13. The basic steps involved in transforming an ingot into metal products.

American Iron and Steel Institute

Inland Steel Company

3-14. Typical blooming mill in operation. Hot ingot (*at right) is ready for rolling once the mill discharges the bloom now being rolled.

involved in transforming a heated ingot into finished mill products. Rolling of metals can involve a variety of steps, depending on what the finished mill product is to be. Typically the first step would be to process the ingots through a blooming mill or a slabbing mill, then possibly through a billet mill. These operations quickly and efficiently reduce the hot ingots to semifinished products called *blooms, slabs,* and *billets.* (These terms will be explained in the following paragraphs.) These processes are basically the same for nonferrous as for ferrous materials.

Rolling Structural Shapes

Rolling of blooms—semifinished steel pieces about $9 \times 9''$ in cross-sectional area—is the first step in the rolling of structural shapes as angles and beams. The bloom is derived from the ingot by rolling it in the *blooming mill.* Fig. 3-14. Most blooming mills are *two-high reversing* mills. The term "two-high" means they have two heavy grooved rolls placed one above the other and revolving at the same speed but in opposite directions. "Reversing" means that the ingots make several passes through the mill in alternating directions. In other words, the ingot is rolled back and forth in the mill. Blooming mills are large and powerful, capable of reducing an average-size ingot ($25 \times 27''$ in cross section) in about 15 passes and less than 5 minutes. The completed bloom is sheared into shorter sections which are

3-15a. Rolling of a wide flange "H" structural beam. Here the bloom is passing through the rolls, developing the beam characteristics.

3-15b. The finished beam is passing through the last stand of the mill.

further rolled into structural shapes in finishing mills. Fig. 3-15 illustrates the rolling of a wide-flange "H" structural beam.

Rolling Plates and Sheets

Rolling of slabs—semifinished steel pieces, rectangular in cross section and of varied sizes—is the first step in rolling plates and sheets. Although small slabs can be rolled in a blooming mill, slabs used for rolling wide plates and sheets are rolled in slabbing mills. *Slabbing mills* are similar to blooming mills but are much wider and have two vertical rolls which control the width of the slab. After the slab reaches the desired size, it is sheared into shorter sections which are further rolled into plates and sheets in finishing mills. The rolling of sheets is a continuous process, with the hot slab entering one side of the continuous mill and being progressively reduced to a long strip of sheet metal coiled at the other end of the mill. Fig. 3-16.

Rolling Bars, Rods, and Wire

Billets—semifinished steel pieces about 2" square in minimum cross sectional area—are rolled in *billet mills* from blooms and are used in bar and wire mills to roll bars, rods, and wires. Billet mills are mostly continuous (not reversing) mills of high volume production, capable of producing many sizes of billets as required in the bar and wire mills. The steel enters one side of the mill as a hot billet and is gradually reduced, leaving the other side as a finished mill product.

Continuous Casting Process

It has been pointed out that molten steel can be used directly in a continuous casting installation to cast blooms, slabs, or billets. Therefore, in a steel mill with

93

3-16. Slabs turned into coils of sheet metal after passing through the continuous strip mill.

Inland Steel Company

a continuous casting installation, there is no need for casting ingots or for installing the expensive slabbing, blooming, and billet mills. Continuous casting came into fairly wide use around 1950 and is growing in importance as more improvements on the process are being made. Continuous casting of blooms, slabs, and billets is less expensive, more efficient, and improves certain properties of steel. The process is relatively simple. Molten steel from the furnace is poured with a ladle into a reservoir (called a *tundish*) above the mold. Fig. 3-17. As the steel flows into the water-cooled mold, it is chilled and develops a shell identical to the shape of the mold. The shell is gradually pulled down by rolls; then as it cools it is straightened and cut into appropriate lengths. Blooms, slabs, and billets from a continuous casting mill are further reduced to finished products by rolling.

Types of Steel, Composition, Properties, and Uses

The great variety of steels produced today almost defies a systematic and detailed description. Often the categories of *carbon steels, alloy steel,* and *special steels* are used. (Each of these groups will be discussed in detail on the following pages.) This classification method is satisfactory for many simple purposes. However, the Society of Automotive Engineers (SAE) developed a more detailed system for designating steels. This system was later adopted by the American Iron and Steel Institute (AISI) and is now widely used in the United States. The SAE-AISI system, which is based on the composition

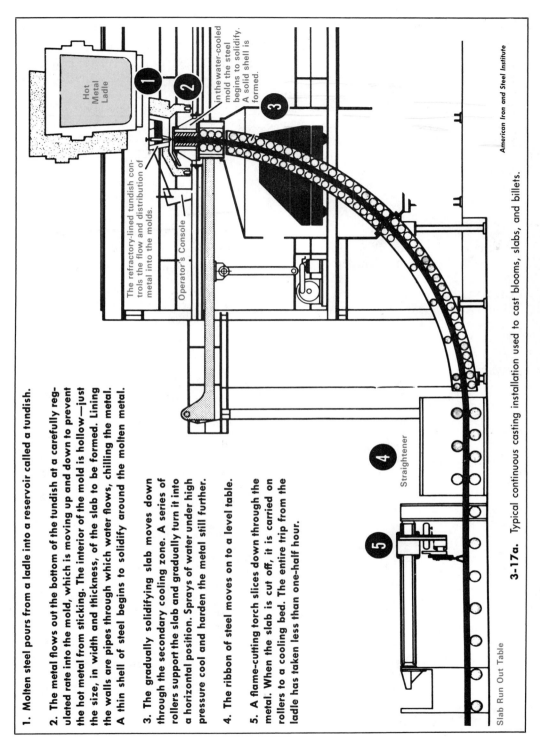

1. Molten steel pours from a ladle into a reservoir called a tundish.

2. The metal flows out the bottom of the tundish at a carefully regulated rate into the mold, which is moving up and down to prevent the hot metal from sticking. The interior of the mold is hollow—just the size, in width and thickness, of the slab to be formed. Lining the walls are pipes through which water flows, chilling the metal. A thin shell of steel begins to solidify around the molten metal.

3. The gradually solidifying slab moves down through the secondary cooling zone. A series of rollers support the slab and gradually turn it into a horizontal position. Sprays of water under high pressure cool and harden the metal still further.

4. The ribbon of steel moves on to a level table.

5. A flame-cutting torch slices down through the metal. When the slab is cut off, it is carried on rollers to a cooling bed. The entire trip from the ladle has taken less than one-half hour.

3-17a. Typical continuous casting installation used to cast blooms, slabs, and billets.

Technology of Industrial Materials

of the steel, is usually stated as a four-digit number. Table 3-B gives the essentials of this system. The first digit denotes the principal alloy element, the second digit indicates the approximate alloy content, and the last two digits indicate the carbon content. Thus *SAE-AISI 1030* indicates that the steel is plain carbon steel containing about 0.30% carbon. *SAE-AISI 2540* denotes a nickel-alloy steel containing about 5% nickel and about 0.40% carbon.

Carbon Steels

Carbon steel is an alloy of iron and carbon, without significant amounts of other elements. Therefore the carbon content plays the most important role in determining the properties of carbon steel. Fig. 3-18. More technically, according to AISI a carbon steel is any steel for which no minimum content is specified for any element added to obtain desirable alloying effects.

About 85% of all steel is carbon steel. About 130 different grades of carbon steel are produced today to meet the growing needs of modern technology. Carbon steel

Table 3-B
SAE-AISI alloy steel designation system.

SAE-AISI Number	Alloy Elements and Approximate Percentages
	Carbon Steels
10xx	Plain carbon steels (0.05–0.90C)
11xx	Free-cutting carbon steels
	Manganese Steels
13xx	(1.75 Mn)
	Nickel Steels
23xx	(3.50 Ni)
25xx	(5.00 Ni)
	Nickel-Chromium Steels
31xx	(1.25 Ni and 0.65 Cr)
33xx	(3.50 Ni and 1.57 Cr)
303xx	(Corrosion and heat resisting)
	Molybdenum Steels
40xx	(Carbon-molybdenum; 0.25 Mo)
41xx	(Chromium-molybdenum; 0.95 Cr)
	Nickel-Chromium-Molybdenum Steels
43xx	(1.82 Ni; 0.50 Cr; and 0.25 Mo)
47xx	(1.05 Ni; 0.45 Cr; and 0.20 Mo)
86xx	(0.55 Ni; 0.50 Cr; and 0.20 Mo)
87xx	(0.55 Ni; 0.50 Cr; and 0.25 Mo)
93xx	(3.25 Ni; 1.20 Cr; and 0.12 Mo)
98xx	(1.00 Ni; 0.80 Cr; and 0.25 Mo)
	Nickel-Molybdenum Steels
46xx	(1.57 Ni and 0.20 Mo)
48xx	(3.50 Ni and 0.25 Mo)
	Chromium Steels
50xx	(Low chromium: 0.27–0.50 Cr)
51xx	(Low chromium: 0.80–1.05 Cr)
51xxx	(Medium chromium: 1.02 Cr)
52xxx	(High chromium: 1.45 Cr)
514xx	(Corrosion and heat resisting)
	Chromium-Vanadium Steels
61xx	(0.95 Cr and 0.15 V)
	Silicon-Manganese Steels
92xx	(0.65–0.87 Mn and 0.85–2.00 Si)
xxBxx	Boron Steels
xxLxx	Leaded Steels

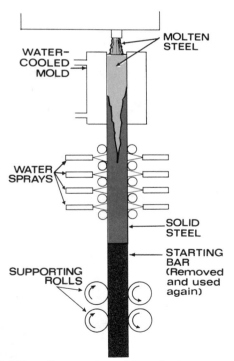

3-17b. Close-up view of Step 2 of continuous casting operation as shown on Fig. 3-17a.

in the hot-rolled form has the lowest cost per pound of all steels and can be used economically in many applications. Carbon-steel mill products cover a wide range of composition, properties, and applications. Table 3-C gives the composition of most common grades of carbon steels. Table 3-D gives the composition of selected free-cutting carbon steels. (These are steels which, because of their composition, are relatively easy to machine and are therefore used for applications requiring a great deal of machining.) Most grades of carbon steel are made into both cold- and hot-rolled products, with cold-rolled still the more expensive. Mill products of carbon steel are produced in wire, rod, bar, sheet, strip, plate, tube, pipe, and structural shapes.

The properties of carbon steel vary with the carbon content. The tensile strength of common grades may vary from 40,000 to 80,000 psi. Table 3-E, page 100, gives estimated mechanical properties and machinability for hot-rolled, cold-finished carbon steel bars. Carbon steels are used in more applications than any other metallic materials. Table 3-F, page 101, gives selected typical applications of carbon steels. Carbon steels can be classified on the basis of carbon content as low-, medium-, and high-carbon steels.

Low-carbon steels contain between 0.08 and 0.35% carbon. In terms of tonnage produced, these steels constitute the most important group. Because of relatively low carbon content, these steels are not used for hardening. However, they are extensively used for structural members as in buildings, bridges, and ships. These steels can be easily welded, formed and forged, but have poor machining properties.

Medium-carbon steels are those with a carbon content between 0.35 and 0.50%. Because of relatively high carbon content, these steels can be hardened by water

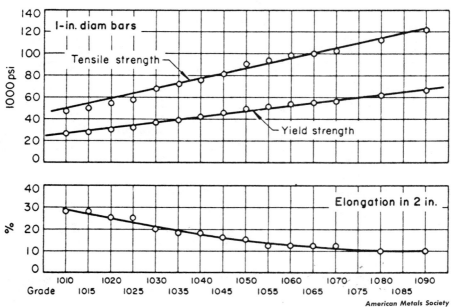

3-18. Effect of carbon content on selected properties of hot-rolled steel.

Table 3-C
Chemical composition limits of selected carbon steels.

SAE-AISI No.	Used for Semifinished Products as Forgings, Hot-Rolled and Cold-Rolled Finished Bars, Wire Rods and Seamless Tubing				Used for Structural Shapes, Plates, Strip, Sheets and Welded Tubing			
	Carbon C	Manganese Mn	Phosphorus P, max	Sulfur S, max	Carbon C	Manganese Mn	Phosphorus P, max	Sulfur S, max.
1006	0.08 max	0.25–0.40	0.04	0.05	0.08 max	0.25–0.45	0.04	0.05
1008	0.10 max	0.25–0.50	0.04	0.05	0.10 max	0.25–0.50	0.04	0.05
1009	0.15 max	0.60 max	0.04	0.05	0.15 max	0.60 max	0.04	0.05
1010	0.08–0.13	0.30–0.60	0.04	0.05	0.08–0.13	0.30–0.60	0.04	0.05
1012	0.10–0.15	0.30–0.60	0.04	0.05	0.10–0.15	0.30–0.60	0.04	0.05
1015	0.13–0.18	0.30–0.60	0.04	0.05	0.12–0.19	0.30–0.60	0.04	0.05
1016	0.13–0.18	0.60–0.90	0.04	0.05	0.12–0.19	0.60–0.90	0.04	0.05
1017	0.15–0.20	0.30–0.60	0.04	0.05	0.14–0.21	0.30–0.60	0.04	0.05
1018	0.15–0.20	0.60–0.90	0.04	0.05	0.14–0.21	0.60–0.90	0.04	0.05
1019	0.15–0.20	0.70–1.00	0.04	0.05	0.14–0.21	0.70–1.00	0.04	0.05
1020	0.18–0.23	0.30–0.60	0.04	0.05	0.17–0.24	0.30–0.60	0.04	0.05
1021	0.18–0.23	0.60–0.90	0.04	0.05	0.17–0.24	0.60–0.90	0.04	0.05
1022	0.18–0.23	0.70–1.00	0.04	0.05	0.17–0.24	0.70–1.00	0.04	0.05
1023	0.20–0.25	0.30–0.60	0.04	0.05	0.19–0.26	0.30–0.60	0.04	0.05
1024	0.19–0.25	1.35–1.65	0.04	0.05	0.18–0.26	1.30–1.65	0.04	0.05
1025	0.22–0.28	0.30–0.60	0.04	0.05	0.21–0.28	0.30–0.60	0.04	0.05
1026	0.22–0.28	0.60–0.90	0.04	0.05	0.21–0.28	0.60–0.90	0.04	0.05
1027	0.22–0.29	1.20–1.50	0.04	0.05	0.21–0.29	1.20–1.55	0.04	0.05
1030	0.28–0.34	0.60–0.90	0.04	0.05	0.27–0.35	0.60–0.90	0.04	0.05
1033	0.30–0.36	0.70–1.00	0.04	0.05	0.29–0.37	0.70–1.00	0.04	0.05
1035	0.32–0.38	0.60–0.90	0.04	0.05	0.31–0.39	0.60–0.90	0.04	0.05
1036	0.30–0.37	1.20–1.50	0.04	0.05	0.29–0.38	1.20–1.55	0.04	0.05
1037	0.32–0.38	0.70–1.00	0.04	0.05	0.31–0.39	0.70–1.00	0.04	0.05
1038	0.35–0.42	0.60–0.90	0.04	0.05	0.34–0.43	0.60–0.90	0.04	0.05
1039	0.37–0.44	0.70–1.00	0.04	0.05	0.36–0.45	0.70–1.00	0.04	0.05
1040	0.37–0.44	0.60–0.90	0.04	0.05	0.36–0.45	0.60–0.90	0.04	0.05
1041	0.36–0.44	1.35–1.65	0.04	0.05	0.35–0.45	1.30–1.65	0.04	0.05
1042	0.40–0.47	0.60–0.90	0.04	0.05	0.39–0.48	0.60–0.90	0.04	0.05
1043	0.40–0.47	0.70–1.00	0.04	0.05	0.39–0.48	0.70–1.00	0.04	0.05
1045	0.43–0.50	0.60–0.90	0.04	0.05	0.42–0.51	0.60–0.90	0.04	0.05
1046	0.43–0.50	0.70–1.00	0.04	0.05	0.42–0.51	0.70–1.00	0.04	0.05
1048	0.44–0.52	1.10–1.40	0.04	0.05	0.43–0.53	1.05–1.40	0.04	0.05
1049	0.46–0.53	0.60–0.90	0.04	0.05	0.45–0.54	0.60–0.90	0.04	0.05
1050	0.48–0.55	0.60–0.90	0.04	0.05	0.47–0.56	0.60–0.90	0.04	0.05
1052	0.47–0.55	1.20–1.50	0.04	0.05	0.46–0.56	1.20–1.55	0.04	0.05
1055	0.50–0.60	0.60–0.90	0.04	0.05	0.50–0.60	0.60–0.90	0.04	0.05
1060	0.55–0.65	0.60–0.90	0.04	0.05	0.54–0.66	0.60–0.90	0.04	0.05
1064	0.60–0.70	0.50–0.80	0.04	0.05	0.59–0.71	0.50–0.80	0.04	0.05
1065	0.60–0.70	0.60–0.90	0.04	0.05	0.59–0.71	0.60–0.90	0.04	0.05
1070	0.65–0.75	0.60–0.90	0.04	0.05	0.64–0.76	0.60–0.90	0.04	0.05
1074	0.70–0.80	0.50–0.80	0.04	0.05	0.68–0.80	0.50–0.80	0.04	0.05
1078	0.72–0.85	0.30–0.60	0.04	0.05	0.71–0.86	0.30–0.60	0.04	0.05
1080	0.75–0.88	0.60–0.90	0.04	0.05	0.74–0.89	0.60–0.90	0.04	0.05
1084	0.80–0.93	0.60–0.90	0.04	0.05	0.79–0.94	0.60–0.90	0.04	0.05
1085	0.80–0.93	0.70–1.00	0.04	0.05	0.79–0.94	0.70–1.00	0.04	0.05
1086	0.80–0.93	0.30–0.50	0.04	0.05	0.79–0.84	0.30–0.50	0.04	0.05
1090	0.85–0.98	0.60–0.90	0.04	0.05	0.84–0.99	0.60–0.90	0.04	0.05
1095	0.90–1.03	0.30–0.50	0.04	0.05	0.89–1.04	0.30–0.50	0.04	0.05

Where silicon is required, the following limits and ranges are commonly used for basic open-hearth steel grades: for steel designations up to but excluding SAE 1015, 0.10% max; for SAE 1015 to SAE 1025, 0.10% max or ranges of 0.10 to 0.20% or 0.15 to 0.30%; for over SAE 1025, ranges of 0.10 to 0.20% or 0.15 to 0.30%.

American Society for Metals

Table 3-D
Chemical composition limits of selected free-cutting carbon steels.

SAE-AISI No.	Carbon C	Manganese Mn	Phosphorus P, max	Sulfur S
1108	0.08-0.12	0.50-0.80	0.040	0.08-0.13
1109	0.08-0.13	0.60-0.90	0.040	0.08-0.13
1115	0.13-0.18	0.60-0.90	0.040	0.08-0.13
1117	0.14-0.20	1.00-1.30	0.040	0.08-0.13
1118	0.14-0.20	1.30-1.60	0.040	0.08-0.13
1119	0.14-0.20	1.00-1.30	0.040	0.24-0.33
1120	0.18-0.23	0.70-1.00	0.040	0.08-0.13
1126	0.23-0.29	0.70-1.00	0.040	0.08-0.13
1132	0.27-0.34	1.35-1.65	0.040	0.08-0.13
1137	0.32-0.39	1.35-1.65	0.040	0.08-0.13
1138	0.34-0.40	0.70-1.00	0.040	0.08-0.13
1139	0.35-0.43	1.35-1.65	0.040	0.12-0.20
1140	0.37-0.44	0.70-1.00	0.040	0.08-0.13
1141	0.37-0.45	1.35-1.65	0.040	0.08-0.13
1144	0.40-0.48	1.35-1.65	0.040	0.24-0.33
1145	0.42-0.49	0.70-1.00	0.040	0.04-0.07
1146	0.42-0.49	0.70-1.00	0.040	0.08-0.13
1151	0.48-0.55	0.70-1.00	0.040	0.08-0.13
1211	0.13 max	0.60-0.90	0.07-0.12	0.08-0.15
1212	0.13 max	0.70-1.00	0.07-0.12	0.16-0.23
1213	0.13 max	0.70-1.00	0.07-0.12	0.24-0.33

When silicon is required, the following limits and ranges are commonly used for basic open-hearth or basic electric steel grades: for steels SAE 1108 and 1109, 0.10% max; for SAE 1115 and over, 0.10% max; 0.10 to 0.20%, or 0.15 to 0.30%. Silicon is not commonly specified for grades up to and excluding SAE 1115.

Acid Bessemer steels are not furnished with specified silicon content. Also, basic open-hearth and basic electric steel equivalents are customarily furnished without specified silicon content.

quench and tempered. However, they should be normalized or annealed before hardening to attain the best mechanical properties. These steels are generally used for forging, high-strength casting, crankshafts, machinery parts, and many common hand tools. Medium-carbon steels are considered the most versatile of all carbon steels because they can be hardened, easily welded, and machined.

High-carbon steels are those with a carbon content over 0.55%. The outstanding characteristic of these steels is that they can be heat-treated more readily than any other carbon steel. However, because of the high carbon content these steels are relatively difficult to fabricate, more expensive, and therefore more limited in application. High-carbon steels are difficult to machine, form, and weld. They are used for springs, hand tools, cutting tools, and agricultural implements. Table 3-F.

Alloy Steels

According to AISI, alloy steels are those steels in which the content and the number of alloying elements are specified within certain limits. Even though alloy steels constitute a relatively small percentage of all steels, their value in modern technology is of considerable importance.

As has been pointed out in Chapters 1 and 2, alloy steels are produced for these purposes:

◆ Increasing strength and hardness, thus improving the weight-to-strength ratio.
◆ Increasing uniformity of mechanical properties.
◆ Increasing resistance to such environmental factors as high or low temperature, corrosion, and wear.

Alloy steels can be classified as low-alloy, medium-alloy, and high-alloy (or special) steels.

Low-alloy Steels. These are steels with less than 4% alloying content. They are of two kinds:

◆ High-strength low-alloy steels.
◆ High-strength heat-treated construction alloy steels.

The low-alloy steels occupy the same position among alloy steels as low-carbon steels occupy among carbon steels. The primary purpose for developing these steels was to improve the weight-to-strength ratio, especially in movable

Technology of Industrial Materials

Table 3-E
Estimated mechanical properties and machinability of hot-rolled and cold-finished carbon steel bars.

SAE-AISI No.	Type of Processing	Tensile Strength, 1000 psi	Yield Strength, 1000 psi	Elongation in 2 in., %	Reduction in Area, %	Brinell Hardness	Average Machinability Rating*
1006	Hot rolled	43	24	30	55	86	
	Cold drawn	48	41	20	45	95	50
1010	Hot rolled	47	26	28	50	95	
	Cold drawn	53	44	20	40	104	55
1015	Hot rolled	50	27.5	28	50	101	
	Cold drawn	56	47	18	40	111	60
1020	Hot rolled	55	30	25	50	111	
	Cold drawn	61	51	15	40	121	65
1025	Hot rolled	58	32	25	50	116	
	Cold drawn	64	54	15	40	126	65
1030	Hot rolled	68	37.5	20	42	137	
	Cold drawn	76	64	12	35	149	70
1035	Hot rolled	72	39.5	18	40	143	
	Cold drawn	80	67	12	35	163	65
1040	Hot rolled	76	42	18	40	149	
	Cold drawn	85	71	12	35	170	60
1045	Hot rolled	82	45	16	40	163	
	Cold drawn	91	77	12	35	179	55
	ACD (a)	85	73	12	45	170	65
1050	Hot rolled	90	49.5	15	35	179	
	Cold drawn	100	84	10	30	197	45
	ACD (a)	95	80	10	40	189	55
1055	Hot rolled	94	51.5	12	30	192	
	ACD (a)	96	81	10	40	197	55
1060	Hot rolled	98	54	12	30	201	
	SACD (b)	90	70	10	45	183	60
1065	Hot rolled	100	55	12	30	207	
	SACD (b)	92	71	10	45	187	60
1070	Hot rolled	102	56	12	30	212	
	SACD (b)	93	72	10	45	192	55
1080	Hot rolled	112	61.5	10	25	229	
	SACD (b)	98	75	10	40	192	45
1085	Hot rolled	121	66.5	10	25	248	
	SACD (b)	100.5	78	10	40	192	45
1090	Hot rolled	122	67	10	25	248	
	SACD (b)	101	78	10	40	197	45
1111	Hot rolled	55	33	25	45	121	
	Cold drawn	75	58	10	35	163	95
1112	Hot rolled	56	33.5	25	45	121	
	Cold drawn	78	60	10	35	167	100
1115	Hot rolled	55	30	25	50	111	
	Cold drawn	61	51	20	40	121	80
1120	Hot rolled	62	34	23	47	121	
	Cold drawn	69	58	15	40	137	80
1132	Hot rolled	83	45.5	16	40	167	
	Cold drawn	92	77	12	35	183	75
1140	Hot rolled	79	43.5	16	40	156	
	Cold drawn	88	74	12	35	170	70
1145	Hot rolled	85	47	15	40	170	
	Cold drawn	94	80	12	35	187	65
1151	Hot rolled	92	50.5	15	35	187	
	Cold drawn	102	86	10	30	207	65

*Properties listed are given as a matter of general information; they do not form a part of requirements. Machinability ratings are based on 1112 steel assumed as 100% machinable. (a) ACD, annealed cold drawn; (b) SACD, spheroidized annealed cold drawn.

American Society for Metals

Table 3-F
Selected typical applications of carbon steels.

SAE-AISI Number	Typical Applications
	Low-Carbon Steels
1006–1012	Used where soft and plastic steel is needed, as in soft sheet, strip, tubing, pipe, and welding.
1015–1022	Used where soft and tough steel is needed, as in rivets, screws, tubing, wire, rods, structural shapes, and strips.
1023–1032	Used for such parts as pipes, gears, shafts, bars, and structural shapes.
	Medium-Carbon Steels
1035–1040	Used for large sections as in forged parts, shafts, axles, rods, and gears.
1041–1050	Used for heat-treated machine parts, such as shafts, axles, gears, and spring wire.
1052–1055	Used for heavy-duty machine parts, such as gears and forgings.
	High-Carbon Steels
1060–1070	Used where good shock resistance is needed, as in forge dies, rails and set screws.
1074–1080	Used for tools requiring toughness and hardness, such as shear blades, hammers, wrenches, chisels, and cable wire.
1084–1095	Used for cutting tools, such as dies, milling cutters, drills, taps, lathe tools, files, knives, and other woodworking tools.

equipment. By improving this ratio, smaller section sizes could be made with low-alloy steels to carry the same load that much larger section sizes made by low carbon steel carry.

The principal alloying elements in low-alloy steels are carbon, phosphorus, molybdenum, manganese, silicon, copper, chromium, and nickel. Phosphorus, manganese, chromium, and nickel are primarily used to improve strength. Molybdenum is used to improve both strength and hardenability. Copper is added to improve resistance to atmospheric corrosion, and silicon serves as a deoxidizer.

Carbon is the most important alloying element in all alloy steels, and it has the same effects as in carbon steels. Table 3-G gives the effects of the principal alloying elements used in steel. The chemical composition of most low-alloy steels is specified in Table 3-H. Low-alloy steels have a minimum yield of 50,000 psi and a minimum tensile strength of 50,000 psi. They can be easily cold-formed, welded, and machined. Low-alloy steels are available in all standard forms such as sheet, strip, plate, structural shapes, and bar. However, the largest percentage is produced in such products as plates, I-beams, angles, channels, and other structural shapes. The major application for these steels is as stationary structural components and components of movable equipment.

Medium-Alloy Steels. These steels were developed primarily to improve hardenability, for use in situations where carbon steels could not function effectively. Dis-

Technology of Industrial Materials

Table 3-G
Summary of the effects of selected principal alloying elements used in steel.

EFFECTS ON STEEL	Boron (B)	Carbon (C)	Chromium (Cr)	Cobalt (Co)	Columbium (Cb)	Copper (Cu)	Lead (Pb)	Manganese (Mn)	Molybdenum (Mo)	Nickel (Ni)	Silicon (Si)	Sulfur (S)	Titanium (Ti)	Tungsten (W)	Vanadium (V)
Improve:															
Abrasion resistance	X	X	X					X							
Corrosion resistance			X			X				X					
Dioxidizing capability								X			X				
Ductility										X					
Elastic limit			X						X	X					
Electrical and magnetic properties											X				
Fatigue resistance											X				X
Grain structure					X			X					X		X
Hardenability	X	X	X					X	X	X					X
Hardness	X	X	X					X							
High-temperature service properties				X					X	X			X	X	
Impact strength			X												X
Machinability							X					X			
Magnetic properties											X				
Shock resistance			X							X					X
Strength (tensile)	X	X	X		X			X	X	X			X	X	
Toughness			X					X	X	X				X	X
Wear resistance	X	X	X	X				X						X	
Workability			X					X							

tortion, residual stresses, and depth of hardening, which are limiting factors in carbon steels, are eliminated with medium-alloy steels. Therefore medium-alloy steels can be used more readily with complicated shapes and large sections because hardening can be achieved with moderate heat-treating temperatures and procedures. It has been found that for improving hardenability, a combination of alloying elements is more effective than the same amount of a single element. The principal alloying elements in medium alloy steels are nickel, molybdenum, va-

Chapter 3. Metallic Materials

Table 3-J
Classification and compositions of principal types of tool steels.

SAE-AISI Designation	C	Mn	Si or Ni	Cr	V	W	Mo	Co
Water-Hardening Tool Steels								
W1*	0.60 to 1.40	....	....	....	....	....	....	....
W2*	0.60 to 1.40	....	....	....	0.25	....	....	....
W3	0.60 to 1.40	....	....	....	0.50	....	....	....
W4	0.60 to 1.40	....	....	0.25	....	....	....	....
W5	0.60 to 1.40	....	....	0.50	....	....	....	....
W6	0.60 to 1.40	....	....	0.25	0.25	....	....	....
W7	0.60 to 1.40	....	....	0.50	0.20	....	....	....
Shock-Resisting Tool Steels								
S1*	0.50	....	....	1.50	....	2.50	....	....
S2	0.50	....	1.00 Si	....	....	....	0.50	....
S3	0.50	....	....	0.75	....	1.00	....	....
S4	0.50	0.80	2.00 Si	....	....	....	....	....
S5	0.50	0.80	2.00 Si	....	....	....	0.40	....
Oil-Hardening Cold Work Tool Steels								
O1*	0.90	1.00	....	0.50	....	0.50	....	....
O2	0.90	1.60	....	....	....	....	....	....
O6	1.45	....	1.00 Si	....	....	....	0.25	....
O7	1.20	....	....	0.75	....	1.75	0.25	opt....
Air-Hardening Medium-Alloy Cold Work Tool Steels								
A2*	1.00	....	....	5.00	....	....	1.00	....
A4	1.00	2.00	....	1.00	....	....	1.00	....
A5	1.00	3.00	....	1.00	....	....	1.00	....
A6	0.70	2.00	....	1.00	....	....	1.00	....
A7	2.25	....	....	5.25	4.50	....	1.00	....
High-Carbon High-Chromium Cold-Work Tool Steels								
D1	1.00	....	....	12.00	....	....	1.00	....
D2*	1.50	....	....	12.00	....	....	1.00	....
D3*	2.25	....	....	12.00	....	....	....	....
D4*	2.25	....	....	12.00	....	....	1.00	....
D5	1.50	....	....	12.00	....	....	1.00	3.00
D6	2.25	....	1.00 Si	12.00	....	1.00	....	....
D7	2.35	....	....	12.00	4.00	....	1.00	....
Chromium Hot-Work Tool Steels								
H11	0.35	....	....	5.00	0.40	....	1.50	....
H12*	0.35	....	....	5.00	0.40	1.50	1.50	....
H13*	0.35	....	....	5.00	1.00	....	1.50	....
H14	0.40	....	....	5.00	....	5.00	....	....
H15	0.40	....	....	5.00	....	....	5.00	....
H16	0.55	....	....	7.00	....	7.00	....	....
Tungsten Hot-Work Tool Steels								
H20	0.35	....	....	2.00	....	9.00	....	....
H21*	0.35	....	....	3.50	....	9.50	....	....
H22	0.35	....	....	2.00	....	11.00	....	....
H23	0.30	....	....	12.00	....	12.00	....	....
H24	0.45	....	....	3.00	....	15.00	....	....
H25	0.25	....	....	4.00	....	15.00	....	....
H26	0.50	....	....	4.00	1.00	18.00	....	....

Technology of Industrial Materials

Table 3-J (Continued)
Classification and compositions of principle types of tool steels.

SAE-AISI Designation	C	Mn	Si or Ni	Cr	V	W	Mo	Co
Molybdenum Hot-Work Tool Steels								
H41	0.65	....	....	4.00	1.00	1.50	8.00	....
H42	0.60	....	....	4.00	2.00	6.00	5.00	....
H43	0.55	....	....	4.00	2.00	....	8.00	....
Tungsten High-Speed Tool Steels								
T1*	0.70	....	....	4.00	1.00	18.00	....	....
T2	0.85	....	....	4.00	2.00	18.00	....	....
T3	1.05	....	....	4.00	3.00	18.00	....	....
T4	0.75	....	....	4.00	1.00	18.00	....	5.00
T5	0.80	....	....	4.00	2.00	18.00	....	8.00
T7	0.75	....	....	4.00	2.00	14.00	....	....
T8	0.80	....	....	4.00	2.00	14.00	....	5.00
T15	1.50	....	....	4.00	5.00	12.00	....	5.00
Molybdenum High-Speed Tool Steels								
M1*	0.80	....	....	4.00	1.00	1.50	8.50	....
M2*	0.85	....	....	4.00	2.00	6.25	5.00	....
M3*	1.00	....	....	4.00	2.40	6.00	5.00	....
M4	1.30	....	....	4.00	4.00	5.50	4.50	....
M6	0.80	....	....	4.00	1.50	4.00	5.00	12.00
M7	1.00	....	....	4.00	2.00	1.75	8.75	....
M10	0.85	....	....	4.00	2.00	....	8.00	....
M15	1.50	....	....	4.00	5.00	6.50	3.50	5.00
M30	0.80	....	....	4.00	1.25	2.00	8.00	5.00
M33	0.90	....	....	3.75	1.15	1.75	9.50	8.25
M34	0.90	....	....	4.00	2.00	2.00	8.00	8.00
M35	0.80	....	....	4.00	2.00	6.00	5.00	5.00
M36	0.80	....	....	4.00	2.00	6.00	5.00	8.00
Low-Alloy Special-Purpose Tool Steels								
L1	1.00	....	....	1.25	....	....	....	....
L2	0.50 to 1.10	....	....	1.00	0.20	....	....	....
L3	1.00	....	....	1.50	0.20	....	....	....
L4	1.00	0.60	....	1.50	0.20	....	....	....
L5	1.00	1.00	....	1.00	....	....	0.25	....
L6	0.70	....	1.50 Ni	0.75	....	....	0.25	opt....
L7	1.00	0.35	....	1.40	....	....	0.40	....
Carbon-Tungsten Tool Steels								
F1	1.00	....	....	....	....	1.25	....	....
F2	1.25	....	....	....	....	3.50	....	....
F3	1.25	....	....	0.75	....	3.50	....	....
Low-Carbon Mold Steels								
P1	0.10 max	....	....	....	....	....	....	....
P2	0.07 max	....	0.50 Ni	1.25	....	....	0.20	....
P3	0.10 max	....	1.25 Ni	0.60	....	....	....	....
P4	0.07 max	....	....	5.00	....	....	....	....
P5	0.10 max	....	....	2.25	....	....	....	....
P6	0.10	....	3.50 Ni	1.50	0.20	....	....	....
P20	0.30	....	....	0.75	....	....	0.25	....
PPT	0.20	1.20 Al	4.00 Ni	....	....	....	....	....

*Produced by the majority of tool steel producers and stocked in most warehousing districts.

American Society for Metals

Chapter 3. Metallic Materials

Table 3-K
ASM groups of tool steels and their major applications.

Group A—Air-hardening medium-alloy cold work tool steels	For intricate die shapes, thread-rolling dies, and slitters.
Group D—High-carbon high-chromium cold work steels	For long-run blanking and forming dies, thread-rolling dies, brick molds, gages, and abrasion-resisting liners.
Group E—Carbon-tungsten tool steels	For paper-cutting knives, wiredrawing dies, plug gages, forming tools, and brass-cutting tools.
Group H11-H16—Chromium hot work tool steels	For hot die work of all kinds: extrusion dies, die-casting dies, forging dies, mandrels and hot shears.
Group H20-H26—Tungsten hot work steels	For heavy hot-forming dies.
Group H41-H43—Molybdenum hot work steels	For hot-work applications similar to tungsten group.
Group L—Low-alloy special purpose steels	For machine tools, bearings, rollers, clutch plates, high-wear springs, chuck parts.
Group M—Molybdenum high-speed steels	For cutting tools of all kinds.
Group O—Oil-hardening cold-work steels	For short-run cold-forming dies, blanking dies, gages, and cutting tools.
Group P—Low-carbon mold steels	For low-temperature die-casting dies, molds for injection or compression molding of plastics.
Group S—Shock-resisting tool steels	For chisels, rivet sets, hammers, and other tools with impact loading.
Group T—Tungsten high-speed steels	For such cutting tools as tool bits, drills, reamers, taps, broaches, milling cutters, and hobs.
Group W—Water-hardening tool steels	Primarily for embossing, coining, striking, and cold-heading dies.

NONFERROUS METALLIC MATERIALS

Many nonferrous metallic materials are produced today, but only a few are extensively used in industry. Like ferrous metals, most nonferrous metals are used industrially as alloys, but limited amounts of a few nonferrous metals are used industrially in pure form (99.90 purity). Among the nonferrous metals and alloys discussed in this chapter are aluminum, copper, zinc, lead, tin, magnesium, silver, gold, and platinum.

Aluminum

Industrial Production of Aluminum

Aluminum is found in the earth's crust as an ore called *bauxite* which contains over 45% aluminum plus oxygen, water, and other impurities. Bauxite is taken from open-pit mines and transported to a refining plant where it is crushed, dried,

Technology of Industrial Materials

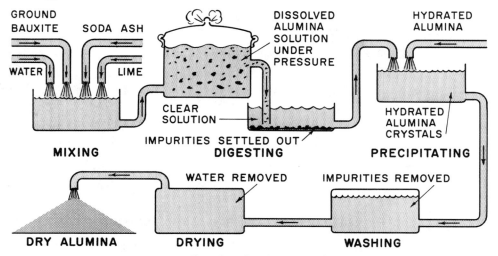

3-19. Flow chart for aluminum refining.

and screened to remove the rough impurities. From this point to the formation of mill products, the aluminum undergoes three major processes:

1. Production of *alumina* (Al_2O_3), a pure aluminum oxide.
2. Production of aluminum ingots.
3. Production of mill products.

The production of alumina involves several steps. First, the finely ground bauxite powder is charged into large tanks (digesters) with caustic soda. This chemically converts the bauxite into sodium aluminate. Fig. 3-19. After the digestion the sodium aluminate is filtered and mixed with a crystalline substance called *alumina hydrate*. This causes the alumina in the sodium aluminate to crystallize and precipitate, yielding larger amounts of alumina hydrate. Next this alumina hydrate is heated in kilns to about 1800° F. This removes the water and leaves a sugarlike powder which is the alumina, ready for melting.

To separate aluminum from oxygen, the powdered alumina is melted in large electrolyte furnaces called *cells*. These electrolytic cells are lined with carbon (cathode), equipped with suspended carbon electrodes (anodes), and filled with electrolyte called *cryolite*. Powdered alumina is charged at the top of the cell and gradually dissolves in the cryolite. Direct electric current of high amperage and low voltage is passed between the anode and the cathode, creating powerful forces which separate the metallic aluminum from the oxygen. The melted aluminum settles to the bottom of the cell, and the oxygen joins with carbon from the anode to form carbon dioxide which escapes from the top of the cell. The melting process is continuous, with new powdered alumina added from the top of the cell and molten aluminum tapped from the bottom in a ladle every few hours. The aluminum is then further treated with fluxes, degassed, and alloyed before it is poured into ingots of various sizes and shapes.

In fabricating aluminum mill products from ingots, the steps are similar to those for fabricating steel mill products from steel ingots. Aluminum mill products are

Chapter 3. Metallic Materials

Table 3-L
Advantageous properties of aluminum.

EASE OF WORKING	Aluminum can be fabricated economically by all the common processes.
WEIGHT	Aluminum weighs far less than many other common industrial metals.
TENSILE STRENGTH	Some aluminum alloys have tensile strength higher than 80,000 pounds per square inch.
CORROSION RESISTANCE	Aluminum needs no protection in most ordinary environments.
COATABILITY	Oxide coatings of many colors and hard, wear-resisting surface finishes can be applied.
ELECTRICAL PROPERTIES	Pound for pound, aluminum has twice the conductance of copper. For equal sections the conductivity of aluminum is 62% that of copper.
MAGNETIC PROPERTIES	Aluminum is nonmagnetic. Thus electrical losses and disturbances are reduced in applications such as cable shielding and electronic equipment.
HEAT CONDUCTION	Because it transmits heat rapidly and efficiently, aluminum is widely used for kitchenware, automotive pistons, industrial equipment, and similar products.
REFLECTIVE PROPERTIES	Aluminum reflects both light and heat with high efficiency.
LOW ELASTICITY	A low modulus of elasticity gives aluminum the ability to withstand considerable impact without deforming permanently.
MISCELLANEOUS	Aluminum is nontoxic and odor-free; thus it is widely used in food-processing plants and home kitchens. It is nonsparking, so it is safe near inflammable and explosive substances.

Aluminum Company of America

produced in standard plate, sheet, strip, foil, tube, structural shapes, wire, rod, bar, and forgings.

Properties and Uses of Aluminum

In less than one hundred years, aluminum has grown from a laboratory curiosity to one of the most important non-ferrous metals. Among the outstanding properties of aluminum are its light weight, strength, workability, conduction of heat and electricity, corrosion resistance, its nonmagnetic and nontoxic characteristics, its ability to reflect light and heat, and its pleasing appearance. Table 3-L. Wrought commercially pure aluminum has yield and tensile strength of 5,000 and 13,000 psi, respectively. It is very soft, with about 45% elongation. Its wear resistance and its fatigue and creep strength are poor. However, with cold-working, alloying, and heat treatment, its tensile strength can be increased to about 82,000 psi, with corresponding increase in wear resistance and in creep and fatigue strength. Aluminum can be hardened by precipitation hardening to improve strength and toughness. Additional information regarding aluminum alloys can be found on Table 3-M.

Technology of Industrial Materials

Table 3-M
Characteristics of wrought aluminum alloys.

Alloy	Usual Commercial Tempers	Sheet and Plate	Rod	Bar	Rolled Shapes	Extruded Shapes	Extruded Tube	Drawn Tube	Pipe	Forgings
EC	O, H12, H13, H17	. .	√	√	②	. .	. .	. .	. .	. .
1160	O, F, H112, H12, H14	√	. .	. .	. .	. .	. .	. .	. .	. .
1100	O, F, H112, H12, H14, H16, H18	√	√	√	. .	. .	. .	. .	. .	. .
3003	O, F, H112, H12, H14, H16, H18	√	. .	. .	. .	. .	√	√	√	. .
Alclad 3003	O, F, H112, H12, H14, H16, H18	√	. .	. .	. .	. .	. .	③	. .	. .
3004	O, F, H112, H32, H34, H36, H38	√	. .	. .	. .	. .	. .	. .	. .	. .
Alclad 3004	O, F, H112, H32, H34, H36, H38	√	. .	. .	. .	. .	. .	. .	. .	. .
2011	T3, T8	. .	√	. .	. .	. .	. .	. .	. .	. .
2014	T4, T6	. .	√	√	√	√	√	. .	. .	√
Alclad 2014	T3, T4, T6	√	. .	. .	. .	. .	. .	. .	. .	. .
2017	T4	. .	√	√	. .	. .	. .	. .	. .	. .
2018	T61	. .	. .	. .	. .	. .	. .	. .	. .	√
2218	T72	. .	. .	. .	. .	. .	. .	. .	. .	√
2024	T3, T36, T4	√	√	√	. .	√	√	√	. .	. .
Alclad 2024	T3, T36, T4, T81, T86	√	. .	. .	. .	. .	. .	. .	. .	. .
4032	T6	. .	. .	. .	. .	. .	. .	. .	. .	√
5050	O, F, H112, H32, H34, H36, H38	√	. .	. .	. .	. .	. .	⑤	. .	. .
6151	T6	. .	. .	. .	. .	. .	. .	. .	. .	√
5052	O, F, H112, H32, H34, H36, H38	√	√	√	. .	. .	. .	√	. .	. .
6061	T4, T6, T62	√	√	√	√	√	√	√	√	√
6062	T4, T6, T62	. .	. .	. .	. .	√	√	√	. .	. .
6063	T42, T5, T6, T83, T831, T832	. .	. .	. .	. .	√	√	√	√	. .
7075	T6	√	√	√	. .	√	√	. .	. .	√
Alclad 7075	T6	√	. .	. .	. .	. .	. .	. .	. .	. .

① Not all products are available in all tempers.
② Channel bus only.
③ Drawn tubes and pipe available with alclad coating on inside only.
④ Hexagon only.
⑤ Annealed temper only.

Chapter 3. Metallic Materials

Table 3-M (Continued)
Characteristics of wrought aluminum alloys

	Strength	Workability (Cold)	Resistance to Corrosion	Weldability (Arc)	Machinability	Electrical Conductivity	Forgeability	Hardness	Typical Uses
....	D,C	A	A	A	B	A	....	D	Electrical conductors.
....	D	A	A	A	B	A	....	D	Chemical equipment and railroad tank cars, for liquids such as hydrogen peroxide.
....	D,C	A,B	A	A	B	B	A	D,C	Sheet metal work, spun hollow ware and decorative parts.
....	D,C	A,C	A	A	B	C	A	D,C	Cooking utensils, chemical equipment, pressure vessels, storage tanks, piping, architectural applications and builders' hardware.
....	D,C	A,C	A	A	B	C	....	....	Tea kettles, chemical equipment, heat exchanger tubes, pressure vessels, storage tanks and flashing.
....	C	A,C	A	A	B	C	....	C	Special sheet metal work, shoe eyelets, screw shells pressure vessels and storage tanks.
....	C	A,C	A	A	B	C	....	....	Industrial building sheet and flashing.
....	B	C,D	C⑥	D	A	C	....	B	Screw machine products.
....	A	C,D	C⑥	B	A	C	C	B,A	Heavy-duty structures, aircraft structures and truck frames.
....	A	C,D	A	B	A	C	....	....	Heavy-duty structures and aircraft structures.
....	A	C	C⑥	D	A	D	....	B	Screw machine products.
....	A	....	C	B	A	C	C	B	Aircraft engine cylinder heads and pistons.
....	B	....	C	B	A	C	D	B	Aircraft engine cylinder heads and pistons, jet engine impellers and compressor rings.
....	A	C,D	C⑥	C	A	D	....	A	Aircraft structures, truck wheels and screw machine products.
....	A	C,D	A	C	A	D	....	....	Aircraft structures and truck bodies.
....	B	....	C	B	C	D	C	B	Pistons.
....	C	A,C	A	A	B	B	....	D,C	Decorative refrigerator parts, builders' hardware and coiled tubes.
....	B	....	B	A	B	C	A	B	Intricate forgings for machine and automotive parts.
....	C	A,C	A	A	B	D	....	C,B	Sheet metal work, home appliances, bus, truck and marine applications, hydraulic tubes, street light standards, clock plates and parts.
....	C,B	B	A	A	B	C	....	C,B	Heavy-duty structures where resistance to corrosion is necessary, bridge railings, marine applications, furniture, truck bodies, truck frames, trailers, railroad cars, tread plate, pipe and pipe flanges.
....	C,B	B	A	A	B	C	....	C,B	Alternate for 6061 alloy.
....	C,B	B,C	A	A	B	B	....	C,B	Pipe, railings, builders' hardware, windows, store fronts and architectural applications.
....	A	D	C	C	A	D	D	A	Aircraft structures and keys.
....	A	D	A	C	A	D	....	....	Aircraft structures.

Aluminum Company of America

⑥ In thicknesses of about ⅛ in. and over, these alloys in the —T3, —T4 and —T36 tempers have a "D" rating.
⑦ A, B, C, and D are arbitrary relative ratings in decreasing order of merit. Where two letters are shown, that on the left is for the softest temper listed and that on the right for the hardest temper listed.

Technology of Industrial Materials

Aluminum has about twice the electrical conductivity of copper for equal length and weight and is used extensively for electrical conductors in wire and bus bar forms. Due to its high corrosion resistance, it is used in such applications as chemical equipment and trim for automotive equipment and architectural components. It needs no protection in ordinary environments. As a good conductor, nontoxic aluminum is used for household appliances, cooking utensils, and food processing equipment. Because aluminum is relatively light, strong, and workable, it is used in applications where weight-to-strength ratio is critical, such as in aircraft, automotive, railway, and shipbuilding industries. Aluminum reflects light and can take anodic coatings (oxide coatings) of many different colors. Therefore it is extensively used in architectural, ornamental, and structural applications, interior automotive trim, trim for appliances, and reflecting signs. Table 3-L.

Effects of Alloying Elements

The characteristics of aluminum can be improved by a combination of alloying, heat treatment, and cold-working.* The effects of the principal alloying elements—copper, manganese, silicon, magnesium and zinc—are of extreme importance. These alloying elements are added in amounts from 1 to 20% to improve casting characteristics, workability, hardenability, machinability, strength, and corrosion resistance. Copper is added to improve hardenability, strength, workability, and resistance to corrosion and to reduce shrinkage in cast aluminum. However, copper tends to reduce elongation and increases the cost. Manganese is added to improve corrosion resistance and strength. Magnesium is added to improve machining characteristics, but it presents problems due to fast oxidation. Silicon improves fluidity and gas tightness, and it reduces shrinkage in cast aluminum. Zinc is used for high strength aluminum alloys, and it improves machining and casting characteristics.

Designation of Aluminum Alloys

There are at least five aluminum designation systems in use today—ASTM, SAE, AMS, Government, and Aluminum Association (AA). However, the four-digit system developed by the AA is the most widely used. In this system, Table 3-N, the first digit denotes the alloy group or major alloying element, the second digit indicates the variation of alloy composition, and the last two digits identify the alloy or aluminum purity.

All compositions in the 1000 (1xxx) series have more than 99.00% aluminum; thus in this series the last two digits indicate the aluminum purity in hundredths of one percent. The second digit indicates modification of impurity level. If the number of the second digit is 1-9, it indicates special control of one or more impurities, with zero indicating no special control of impurities. For example: 1065 designates an aluminum 99.65% pure with no special control on impurities. But 1165-1965 designates aluminum of 99.65% purity with some special control of the impurities involved.

For aluminum designated with number 2XXX-8XXX the last two digits are used to identify the different alloy in each group.

In addition to the four-digit alloy designation, the *temper* of aluminum is also designated with a letter followed by one or more numbers. The letters O, F, H, T, and W are used. They follow the four-

---
*See Appendix B for typical mechanical properties of aluminum alloys.

Table 3-N
Wrought aluminum alloy four-digit designation system.

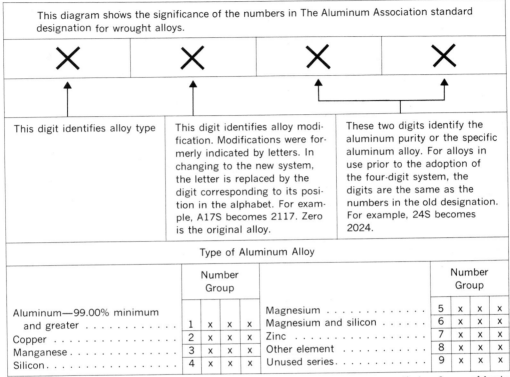

Aluminum Company of America

digit number, separated from it by a hyphen. For example: 1100-O, 1060-H12, 2011-T3, or 1265-F. The letter *O* indicates the annealed temper of wrought aluminum; *T* followed by one or more numbers indicates a heat-treated alloy; *F* indicates the as-fabricated condition of wrought alloys; *H* followed by a number indicates a cold-worked temper; and *W* indicates that the alloy has been quenched from the solution heat treatment.

Copper

Production of Copper

Production of the various types of copper involves four steps. Fig. 3-20:

- Production of matte.
- Production of blister copper.
- Production of electrolytic copper.
- Production of tough pitch copper.

Copper is produced from sulfide ores which are mined either by the shaft or open-pit method, concentrated by various methods, and roasted in special kilns to remove such impurities as arsenic, antimony, and sulfur. The concentrated roasted sulfide ore is then melted in blast or reverberatory furnaces to produce an impure copper called *matte*. The matte cannot be used without further refinement; therefore it is placed in a converter where air is blown through the molten

Technology of Industrial Materials

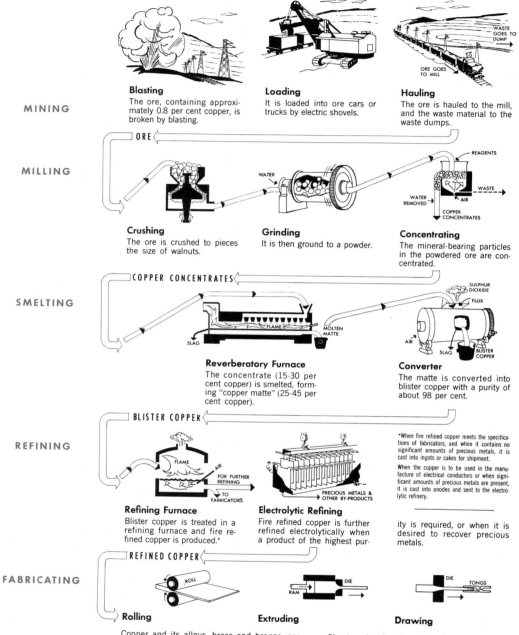

3-20. Flow chart for copper production.

Chapter 3. Metallic Materials

Table 3-Q

ASTM standard four-part system for magnesium alloy designation.

First part	Second part	Third part	Fourth part
Indicates the two principal alloying elements.	Indicates the amounts of the two principal alloying elements.	Distinguishes between different alloys with the same percentages of the two principal alloying elements.	Indicates conditions and properties.
Consists of two code letters representing the two main alloying elements arranged in order of decreasing percentage (or alphabetically if percentages are equal)	Consists of two numbers corresponding to rounded-off percentages of the two main alloying elements and arranged in same order as alloy designations in first part.	Consists of a letter of the alphabet assigned in order as compositions become standard	Consists of a letter followed by a number. (Separated from the third part of the designation by a hyphen)
A—Aluminum B—Bismuth C—Copper D—Cadmium E—Rare Earth F—Iron H—Thorium K—Zirconium L—Beryllium M—Manganese N—Nickel P—Lead Q—Silver R—Chromium S—Silicon T—Tin Z—Zinc	Whole numbers This system is standard for both magnesium and aluminum alloys; thus, designation should be preceded by the name of the base metal unless base metal is obvious; but in this table, only the designations most commonly used in magnesium standards are given.	Letters of alphabet except I and O	F—As fabricated O—Annealed H10 and H11—Slightly strain hardened H23, H24 and H26—Strain hardened and partially annealed T4—Solution heat treated T5—Artificially aged only T6—Solution heat treated and artificially aged

and elongation of 4 to 23%. Table 3-P. Because of poor cold-working characteristics, most magnesium alloys are extruded. However, the light weight of these materials makes possible extremely high weight-to-strength ratios. Thus magnesium alloys are valuable to the aircraft industry and for use in movable equipment, portable tools, and other applications where weight is a critical factor. Most magnesium alloys have good welding and casting characteristics, which make them valuable for the automotive and construction industries. Magnesium alloys are available in ingot form; extruded rod, bar, and structural shapes; strips, sheets, and plates; and tubes and forgings.

Designation of Magnesium Alloys

The designation system most commonly used for magnesium alloys is the four-part system developed by ASTM. Table 3-Q. The first part, consisting of two letters, indicates the two principal alloying elements. The second part, consisting of two numbers representing percentages, indicates the amount of the two principal alloying elements. The third part distinguishes between different alloys which have the same percentages of the two principal alloying elements. The last part indicates certain conditions and properties. For example: AZ63A-T6 indicates a magnesium-aluminum-zinc alloy with 6% aluminum and 3% zinc which has been solution-heat-treated and artificially aged.

Precious Metals

In terms of amount used and dollar value, the three most important precious metals are *gold, silver,* and *platinum.* Although some precious metals can be mined in almost pure form, those most used industrially are recovered during the production of such nonferrous metals as copper, tin, zinc, and lead. Due to the high purity often required for precious metals (99.999% pure), most of them are refined by the electrolytic process. However, the production, refining, and purity of these metals are subject to strict legal requirements in every country. Because of their workability, natural purity, castability, and corrosion resistance, gold and silver have been used for thousands of years and have played an important role in the rise and fall of many civilizations.

Properties and Uses of Gold

Among the important properties of gold are corrosion resistance, good resistance to tarnishing, workability, castability, softness, malleability, and ductility.

The major applications of gold alloys are in the electrical and electronics industry—specifically in such devices as electron tubes, transistors, resistors, potentiometers, and conductors. Other important applications are in the instruments, jewelry, chemical, and dental industries. The jewelry industry uses a sizable amount of gold, either in pure or alloy forms, for jewelry and ornamental novelties. The dental industry uses gold alloys with good mechanical properties for dental fillings. The chemical industry uses gold for lining special chemical equipment to resist corrosion. The principal alloying elements added to gold to improve mechanical properties are silver, copper, and platinum. The purity of gold is usually expressed in *carats.* For example, 12 carats means 50% purity, and 24 carats means 99.999% purity, since absolute purity is practically unattainable.

Properties and Uses of Silver

Among the important properties of silver are resistance to oxidation, thermal and electrical conductivity, high reflectivity, good coating characteristics, formability, photosensitivity, and relatively low melting point. Silver has the highest electrical conductivity of all metals; consequently it is extensively used in the electrical industry for many applications. Because of its photosensitivity, silver is used in photography. Due to its low melting point, it is used in silver solder. Its high reflectivity and resistance to corrosion coupled with good workability make silver valuable for jewelry and tableware.

Properties and Uses of Platinum

Of the precious metals, platinum is the most recent to come into important use. The basic research leading to its refinement was not done until the eighteenth century.

Softness, ductility, and corrosion resistance are among the valuable properties of this metal.

mounted blades, with the bottom blade fixed in position and the top one mounted on a movable ram driven by a motor or hand operated. The workpiece is positioned between the two blades, and the ram is driven downward. When the top blade first contacts the workpiece, the metal is plastically deformed as it is squeezed between the two blades. The blades continue to penetrate the metal until they overcome the strength of the piece. The remaining metal then breaks. *Perforating* is a special kind of shearing operation. A hole is produced in the material by the shearing action of a punch. The punch is the size and shape of the hole to be made and is mounted on the ram of a press. The bottom die on the press contains a hole that mates with the punch. The material to be perforated is placed between the punch and die. The descending punch passes through the material and into the hole in the bottom die. A single piece having the approximate size of the punch cross section is sheared out. Punching, blanking, and nibbling are related shearing operations.

Abrading

Abrading, or grinding, operations remove metal by the action of a rapidly moving abrasive tool pressed against the workpiece. The tool is usually a rotating wheel made up of small, rough grains of an extremely hard substance such as aluminum oxide or silicon carbide. The abrasive particles in the grinding wheel are bonded together with a softer substance such as resin or rubber. Each of the particles on the surface of the wheel acts as a miniature cutting tool and removes a tiny chip of metal. As grinding progresses, the particles on the outside of the wheel become dull and are worn away to expose fresh cutting edges. Grinding produces the same shapes as turning, planing, milling, and other common metal-cutting methods but removes the metal with a grinding wheel instead of a cutting tool.

The principal advantages of grinding are much smoother finished surfaces and more accurate shapes and dimensions on the completed part. Grinding, however, is much slower than other machining methods. Many items are produced by machining the workpiece to approximate size by standard cutting methods and then finishing the part by grinding. Another advantage of grinding is that it can shape extremely hard materials that would be difficult to machine by other methods.

Surface grinding is done to produce flat surfaces. The workpiece is mounted on a flat table that moves under a rotating grinding wheel. Fig. 3-25. Some machines have specially shaped grinding wheels to produce desired contours; others are used for off-hand grinding, such as the sharpening of tool bits. Coated abrasive belts, discs, and drums are also used in metal abrading.

Shaping (Planing)

Shaping and planing are basically the same process. They are done for the same purpose—making straight cuts on flat or slighty curved surfaces. In shaping and the related operations done on planing machines, metal is cut from the surface of the workpiece by a straightforward cutting action. The main difference is that in shaping, the tool reciprocates, whereas in planing it is the workpiece that moves.

Surface grinding, described above, is similar to planing in that the workpiece reciprocates under a stationary grinder.

The planer has a large, flat table that can slide back and forth on the ways or rails of a bed. A stationary housing at the center of the bed holds the cutting tools,

Technology of Industrial Materials

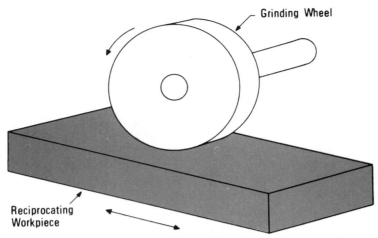

3-25. Abrading is a cutting method involving metal removal by the action of mineral particles.

and the workpiece is fastened to the bed. The workpiece moves into the cutting tool to remove the metal. Planers are used mainly for the machining of heavy-duty parts.

The shaper, Fig. 3-26, is smaller than the planer and somewhat more versatile in operation. It has a reciprocating, horizontal ram mounted over the table of the machine. The ram usually carries only one cutting tool. Shapers have table lengths and cutting strokes ranging from about 6″ up to 3′. The workpiece is clamped to the table and machined as the tool passes back and forth over the surface of the piece. The shaper table moves laterally after each stroke to position uncut material under the tool. Vertical movement of

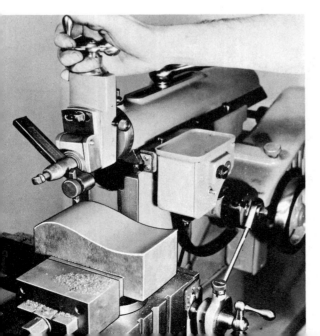

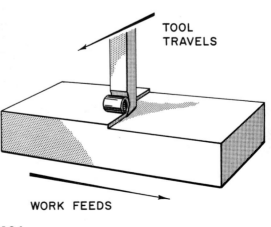

3-26. Typical shaping diagram and machine. Planing is a related operation.

134

Chapter 3. Metallic Materials

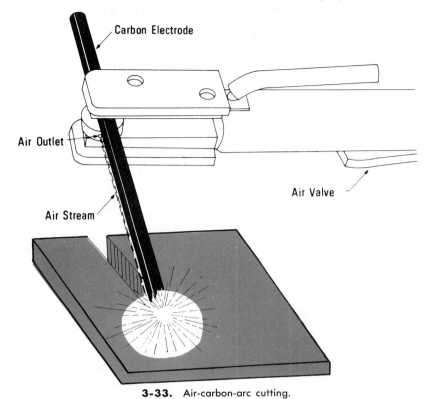

3-33. Air-carbon-arc cutting.

3-34. Three basic metal bending operations: a. Press-brake bending. b. Roll-bending. c. Stationary-die bending.

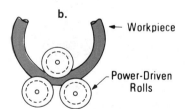

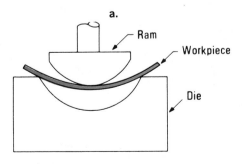

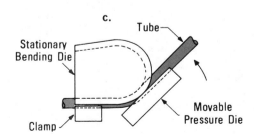

139

Technology of Industrial Materials

ful way of producing many shapes. In contrast with more complex forming operations, bending requires little plastic flow and does not drastically change the workpiece. A bend is usually made along a straight line and results in little or no change in the thickness of the material.

Press-brake bending is the most widely used method of bending materials other than tubing. The workpiece is supported at two points and a force is applied between the points. The press brake is a mechanically or hydraulically powered press with a ram mounted either horizontally or vertically.

Roll bending is done to produce circular bends. The equipment consists of three rolls, with one roll positioned above and between the other two. The three-roll bender is used to bend rod, bar, or tubing into full or partial circles and to bend sheet or plate into cylinders. Wheel-shaped rolls are used to bend narrow sections. Cylindrical rolls are used to bend sheet stock.

Sharp bends in tubing are difficult to make because the hollow tube collapses or flattens at the bend. Some means of supporting the tube from the inside is necessary to make short-radius bends without flattening of tube.

There are several types of equipment for making short-radius bends in tubing. The *stationary-die bending machine* is a common type. It consists of a fixed bending die and a movable wiper block. The tube is wrapped around the die by the wiper block. Stationary-die bending is also called compression bending.

Casting

Metal casting is the production of workpieces by introducing molten metal into a prepared cavity, or mold, and allowing it to solidify. Casting is particularly useful as a means of producing very large or very complex shapes, ranging from automobile engine blocks to intricate machine parts. The process requires that an exact replica, or pattern, of the cast piece be made; from this is made the mold which receives the melted metal.

Sand casting is one of the most widely used techniques. Fig. 3-35. Such castings are likely to have coarse surfaces (because the sand is coarse) and tend to be porous because of gases or impurities trapped when cooling. For many products, such as the hydrants in Fig. 3-36, the coarseness and porosity are not problems, for any mating surfaces are machined to insure a close fit.

When such structural deficiencies would be a problem, they can be overcome by using other casting methods, such as *die casting*. By this method, castings are produced by injecting small quantities of

3-35. Casting diagram, showing sand-and-flask system. Such cast pieces shrink when cooled which requires that the patterns be made slightly oversize.

3-36. The new and old style fire hydrants shown here were made by sand casting.
East Bay Municipal Utility District, Oakland, California

3-37. Die-cast part, showing intricate die used in this process.

metal into cooled, metallic dies (or molds) which solidify quickly. Fig. 3-37. The resultant workpiece has a smoother finish, is less porous, and is not likely to shrink; such characteristics make this a suitable method of producing small, intricate parts.

Other casting methods include *centrifugal casting,* in which molten metal is flung outward to the wall of a fast-rotating, hollow die. Pipes and cylinders of large dimension can be produced in this way.

Investment casting is a technique for making small, complex parts to close tolerances. Also called the "lost wax" process, this involves preparing a wax replica of the workpiece, which is then "invested" in a cylinder of plaster or sprayed with a fine, bonded sand. A subsequent overbaking melts out the wax, which leaves a very accurate mold from which a casting is made. Readers may find this most accurate process familiar, as it is used to make metal fillings and bridgework in dentistry.

Thus casting is important in the production of objects which vary widely in size, complexity, properties, and other characteristics.

Forging

Forging is the most widely used method of hot-forming finished shapes. Rods and bars produced by hot-rolling are frequently formed into the finished article by forging.

Forging is the process of forming hot metal by impact blows (hammering) or by squeezing (pressing). Modern forming methods are basically the same ones used by blacksmiths since ancient times. The metal is heated to a high temperature and then shaped by pressure applied by one surface (the hammer) as the metal rests on another surface (the anvil). Fig. 3-38.

The arm of the blacksmith has been replaced by powerful equipment that can forge intricate shapes from the toughest metals. Forging equipment is of two general types: the *hammer* and the *press.* Both types shape the metal between two dies, one to hold the workpiece and one to deliver the pressure. The dies may be flat surfaces or they may be cut to correspond

141

Technology of Industrial Materials

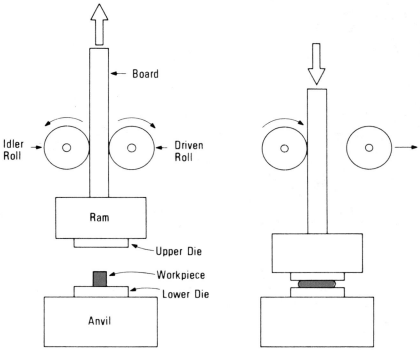

3-38. The board drop hammer is a forging machine which forms workpieces by high-impact blows. Forging also includes squeezing pressures.

to the finished shape of the piece. Fig. 3-39. The hammer delivers high-velocity, impact pressure; the press delivers slower, squeeze pressure. Impact pressure is greatest when the hammer first strikes the workpiece. Squeeze pressure builds up as pressing continues and reaches a maximum just before the pressure is released.

A special form of forging called *cold-heading* is shown in Fig. 3-40. As the name implies, the metal is formed without being heated.

Pressing

Pressing is a process used to form a depression in a flat sheet metal part or to form shallow cup shapes. Drop-hammer forming, a typical pressing operation, is done by the impact blow provided by a falling ram, in contrast to the slow pressing action used for deep drawing.

Drop hammers for cold forming operate the same as hammers for hot forging. A matched set of dies is used, with one die mounted beneath a vertical ram to which the other die is attached.

Forming with the drop hammer is done by placing the sheet stock on the bottom die and then releasing the raised ram. The upper die strikes the workpiece and bends or stretches the metal as necessary to make it conform to the shape of the dies. Forming is usually done in one blow. Parts that require large amounts of deformation are made with two or more sets of dies so that the workpiece can be pro-

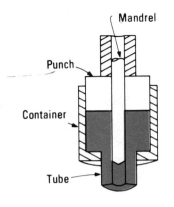
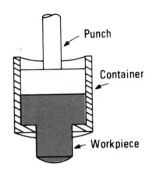

FORWARD EXTRUSION
OF HOLLOW PRODUCT

FORWARD EXTRUSION
OF SOLID PRODUCT

3-46. Basic extrusion methods.

ward extrusion if metal flow is opposite to the motion of the punch. Other directions of metal flow are possible, depending on the location of the hole in the container. Fig. 3-46 shows some typical extrusion methods.

To produce tubular products by forward extrusion, the workpiece is either pre-drilled with a hole for the mandrel or is pierced by the mandrel before being extruded.

Rolling

As explained more fully earlier in this chapter, *cold-rolling* and *hot-rolling* are done by squeezing a workpiece between rotating, cylindrical rolls. Fig. 3-47. Cold-rolling, however, is normally done only to produce thin, flat products such as sheet and strip. Other shapes are more conveniently produced by cold-drawing.

Cold-rolling mills usually have smaller rolls than hot-rolling mills. Cold-rolling requires much more force; the smaller rolls provide this by concentrating the available force over a smaller area. Smaller rolls also enable thinner material to be rolled. Thin material does not absorb all of the rolling force; thus in effect, one roll acts directly against the other. Under such conditions, large rolls tend to flatten at the areas where they contact the workpiece. The thinner the material being formed, the more the rolls tend to become flattened. Thus there is a minimum thickness for material formed by rolling. The minimum varies depending on the diameter of the rolls, but smaller rolls must be used to roll thinner materials. Cold-rolling is seldom done to produce finished products. Cold-rolled sheet and strip are made into finished articles by other methods of fabrication.

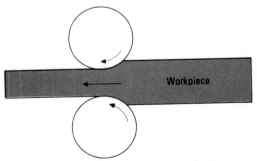

3-47. The operation of the rolling mill. The two rotating rolls squeeze the metal between them to form the new shape.

Technology of Industrial Materials

Spinning

Spinning is done to produce shapes similar to those produced by deep drawing. Spun parts, however, can be made with contoured sides. Many contours that cannot be readily produced by deep drawing can be produced by spinning. The nature of spinning requires that spun parts have shapes that are symmetrical about the axis of rotation. That is, all contours of the shape must be continuous around the article.

Although special equipment is used to spin some articles, especially large parts, most spinning is done on ordinary lathes. The lathe provides the rotating horizontal shaft necessary for the operation. The basic requirements for spinning are shown in Fig. 3-48. As in most metal forming, the workpiece is made to conform to the shape of a die by the application of force. In spinning, the die is called a chuck (or sometimes a mandrel or former). The chuck is attached to the headstock of the lathe. The blank or workpiece, a circle of sheet or strip stock, is clamped at its center against the chuck by the lathe tailstock. The entire assembly rotates in unison. Forming pressure is applied by a roller tool held against the rotating blank.

Metal Fastening

Mechanical Fastening

Mechanical fastening methods are widely used to join metal parts perma-

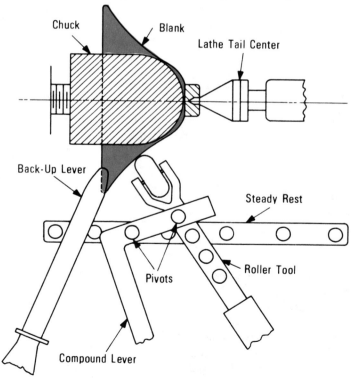

3-48. Top view of metal spinning operation. The metal blank is being formed to the shape of the chuck by the applied pressure of the roller tool.

3-54. Brushing provides a scratch-resistant finish to metal products.

or flame and propels the molten metal in a spray of fine drops to the workpiece surface. Electroplating is a related process.

Protective and beautifying finishing techniques also include brushing, Fig. 3-54, buffing, polishing, and embossing. The first three are actually abrasive processes, while the fourth is achieved by a pressing operation.

Special Metal Processes

The requirements of contemporary technology have placed new demands upon the metalworking industry—for example, developing and working ultra-hard metals, and forming macro and micro parts. Consequently, there have emerged over the years some new processes for working metals. Several classic examples are described below.

Explosive forming is primarily used to make large parts from sheet or plate stock. The process is similar to flexible-die forming, but the force is supplied by detonation of an explosive such as blasting powder, dynamite, or TNT. The greatest advantage of explosive forming is its ability to deliver large amounts of power at low cost.

The most common method of explosive forming is shown in Fig. 3-55. The workpiece is placed over a hollow die in a tank of water. A vacuum line is connected to the die cavity to remove the air from the space between the workpiece and the die. An explosive charge is suspended in the water at a predetermined distance from the workpiece. When the charge is detonated, the explosive force is transmitted to the workpiece, imparting to it a desired shape.

Electrical discharge machining (EDM) is the removal of metal by the energy of an electric spark that arcs between a tool and the surface of the workpiece. The tool and workpiece are immersed in a fluid, such as oil, that will not conduct electricity. The tool is positioned opposite the area of the workpiece to be machined and at a slight distance from it. Rapid pulses of electricity are delivered to the tool, causing sparks to jump from the tool to the work-

Technology of Industrial Materials

piece. The heat from each spark melts away a small amount of metal. As the metal is removed, it is cooled and flushed away by the fluid being circulated through the spark gap. The spark also removes material from the tool, so the tool is slowly consumed as machining progresses. Because of the destruction of the tool, complex shapes are not often produced by EDM. The greatest use of the process is to produce holes or cavities.

Electrochemical machining (ECM) is the removal of metal from a workpiece by dissolving the metal in a chemical solution with the aid of an electric current. Metal removal is controlled by a tool, but the tool does not contact the workpiece. The metal removed is in the form of tiny dust-like particles instead of chips.

The process works on the same principle as electroplating, and the metal removed from the workpiece would normally become plated on the tool. To prevent this, the electrolyte is continuously circulated through the space between the tool and workpiece. Thus the metal is carried away by the electrolyte before it reaches the tool.

Chemical machining (CHM) is the removal of metal by the chemical attack of corrosive liquids such as acids. No tool or electricity is used. The workpiece is submerged in the chemical, and metal is uniformly dissolved from all exposed surfaces. Areas of the workpiece that are not to be machined are covered with a material, such as plastic, that is not affected by the chemical.

Thin material can be completely penetrated, or cut, by CHM. Complex shapes such as electrical circuit boards are separated from thin sheets of material by masking all of the material except the outline of the shape. That procedure is usually called *chemical blanking.* When metal is removed from broad areas of a workpiece, the process is often referred to as *chemical milling,* which is a form of etching.

Laser welding is done with heat supplied by a laser beam, an intense beam of light. The laser beam is produced by focusing flashes of light from high-intensity lamps on a transparent crystal, usually a ruby, which then emits a concentrated beam of light. The light beam is converted to heat when it strikes a nontransparent substance.

Laser welding has many valuable features, but its use is limited by high cost, large power requirements, and slow welding speeds. Most light sources deliver in-

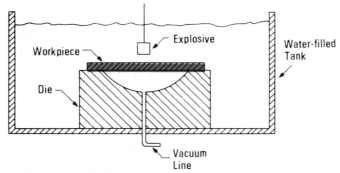

3-55. Explosive forming in a fluid. The explosion causes the workpiece to assume the contour of the die.

termittent flashes, resulting in a pulsed beam. Slow welding speeds are caused by the delay between pulses of the beam.

CONCLUSION—SELECTION OF METALLIC MATERIALS

This chapter has covered the production and processing of metallic materials. This information is essential for the proper selection of a metal for a specific use.

With the constant development of new materials and new areas of application, the selection of metallic materials has become an enormous problem. In fact, the ultimate test of knowledge of materials is at the selection level.

The problem of materials selection was not as critical in the past when very few materials were used and there were not so many advanced technological areas. Today, however, when many materials may seem suitable for a given application, making the best selection requires extensive knowledge of technological and economic requirements of the application as well as the properties of the materials.

Therefore in the selection of materials the two principle factors under consideration are *technology* and *economics*. Although in many situations the ultimate decision will be based on one factor alone —cost—it is of importance that the knowledge of materials technology be used in reducing cost to a minimum. Traditionally, material selection has been based on *cost per unit of strength*.

The material selection process must be guided by the application requirements. In other words, where is the material going to be used? These requirements, derived by careful analysis of the application, serve as primary selection criteria. Some of these requirements would dictate the chemical composition and mechanical properties, form, size, condition, surface finish, processing methods, and quantity required. All these factors play an important role in the material selection process and should be considered on the basis of information presented in this text.

READING LIST

Specific References

For additional information regarding topics discussed in this chapter, the following publications are recommended.

Aluminum Company of America. *Alcoa Aluminum Handbook,* Pittsburgh, PA: Alcoa, 1962.

———. *Alcoa Structural Handbook,* Pittsburgh, PA: Alcoa, 1956.

American Society of Mechanical Engineers. *Materials Technology,* New York: ASME, 1970.

Bain, E. C. and H. W. Paxton. *Alloying Elements in Steel,* Metals Park, OH: American Society for Metals, 1961.

Campbell, I. E. and E. M. Sherwood (Eds.). *High-Temperature Materials and Technology,* New York: John Wiley and Sons, Inc., 1967.

DuMond, T. C. (Ed.). *Engineering Materials Manual,* New York: Reinhold Publishing Corporation, 1951.

Hausner, H. H. and M. G. Bowman (Eds.). *Fundamentals of Refractory Compounds,* New York: Plenum Press, 1968.

Technology of Industrial Materials

Keyser, C. A. *Materials of Engineering*, Englewood Cliffs, NJ: Prentice-Hall, Inc., 1965.

Lyman, T. (Ed.). *Metals Handbook, Volume 1*, Metals Park, OH: American Society for Metals, 1961.

Mantell, C. L. (Ed.). *Engineering Materials Handbook*, New York: McGraw-Hill Book Company, 1958.

McGannon, H. E. (Ed.). *The Making, Shaping and Treating of Steel*, Pittsburgh, PA: United States Steel Corporation, 1964.

Murphy, Glenn. *Properties of Engineering Materials*, Scranton, PA: International Textbook Company, 1957.

Parker, E. R. *Materials Data Book*, New York: McGraw-Hill Book Company, 1967.

Patton, W. J. *Materials in Industry*, Englewood Cliffs, NJ: Prentice-Hall, Inc., 1968.

Samans, C. H. *Metallic Materials in Engineering*, New York: The Macmillan Company, 1963.

Scientific America. *Materials*, San Francisco: W. H. Freeman and Company, 1967.

Sims, C. E. (Ed.). *Electric Furnace Steelmaking*, New York: The American Institute of Mining, Metallurgical and Petroleum Engineers, 1967.

Smith, R. C. *Materials of Construction*, New York: McGraw-Hill Book Company, 1966.

Woldman, N. E. *Engineering Alloys*, New York: Reinhold Publishing Corporation, 1962.

philic plastics exhibit ionic conduction (that is, ions rather than free electrons transmit the charge). Therefore if a plastic is to be used as an insulator outdoors, its resistance to water and humidity must be determined.

In Chapter 1, the principles of a capacitor were discussed. A capacitor requires a dielectric material, one which can store an electric charge and act as an insulator. Plastics are well suited to this application as evidenced by their high dielectric strength.

Thermal Conductivity

Plastics exhibit very low thermal conductivities, which makes them useful as insulators against heat or cold. There are three structurally associated reasons for their characteristic insulating abilities.

- First, the electrons are closely bound by covalent bonds and cannot transmit heat energy as they can in a metal.
- Second, there are large spaces between atoms and molecules of plastic, as indicated by the low density of these materials. Atomic vibrations are not readily passed through such a structure. Thus atomic heat conductivity in plastics is also small.
- Finally the plastic structure voids are insulating areas in themselves. This can be further enhanced when a plastic is processed into a foam, since foam contains many more voids or gas pockets. These pockets add to the insulating qualities of the material.

Density

Plastics exhibit very low densities. Commercial plastics range from 0.30 to 0.75 pounds per cubic inch in density. They are only about one-sixth as dense as steel and half as dense as aluminum. Plastics are ideal where light weight is important.

Corrosion Resistance

Polymeric materials generally exhibit excellent corrosion resistance. As mentioned earlier, the large size of plastic molecules prevents them from going into solution easily. For solution to occur, the solvent must have sufficient spaces in its structure to hold macromolecules. Most solvents and common household liquids do not have structures that can dissolve plastics. However, certain organic solvents with large molecular weights, such as alcohol or gasoline, can readily dissolve some plastics.

Plastics are subject to a special type of corrosion called *swelling*. On account of the low density of plastics, certain solvents—especially organic hydrocarbon compounds such as greases and oils—can enter the intermolecular spaces in the plastic structure. This causes the plastics to swell, leading to a loss in strength.

Prolonged exposure to moisture, oxygen, and sunlight will degrade many polymers. The chemical action of oxygen or water breaks down structural bonds. The energy in sunlight can also cause bonds to break. Polystyrene, for example, is limited in outdoor usage unless painted and kept dry. The sulfur bond in vulcanized rubber can be broken by ozone, a form of oxygen.

The covalent bond characteristic of plastics is usually stable enough to prevent reaction with other materials. It also prevents electrolytic corrosion of most plastics, though some plastics may corrode if they contain metal ions either as a constituent part or from dyes. Although polymers are considered corrosion resistant in general, each application and environment must be examined to insure proper service.

The corrosion resistance of plastics causes pollution problems, since scrap or waste pieces are not degraded easily. As discussed earlier, sunlight, oxygen, and water will break down some plastics. Increased use of such degradable plastics might somewhat reduce the pollution problem. However, in many applications the plastic must resist these elements sufficiently to yield the proper service life. The development of degradable polymeric materials and disposal techniques represents a major challenge to materials technology.

CLASSIFICATION OF PLASTICS

Classification by Composition and Processing

Because plastics were originally used in no-load applications, the early systems of classification stressed processing and composition Table 4-D, pages 166–167. When classified according to processing, plastics are divided into two main groups:

- *Thermoplastics.* These are plastics which can be remelted or softened by heat and then formed while hot. (In industry, frequent remelting is avoided because contamination and some chemical degradation is encountered.) Thermoplastics are supplied by the manufacturer in forms suitable for machining or remelting. Injection molding and extrusion are methods of shaping thermoplastics.
- *Thermosets.* These are plastics which cannot be melted once they have solidified. Thermosets are formed by placing the raw materials, usually called resins, into a mold and then hardening them. They are available as liquids or powders, to be reacted at the user's convenience. In general, thermosets are stronger and harder than thermoplastics. Composition is based upon the smaller organic molecule (monomer) from which the plastic was made.

Classification by Properties and Structures

Recent developments in the plastics industry stimulated the evolution of more comprehensive classification systems based on structural factors and properties. With increasing use of plastics as engineering materials, exact knowledge about their specific properties and behavior was needed to insure satisfactory service under load. While this need was being met, materials science also provided new insights into the structure of plastics. Thus, considerable knowledge was gained about the relationships between structural factors and properties. These relationships serve as the basis for new classification systems. Table 4-E, pages 168–169 gives the key structural facts and properties for many types or classes of plastics used in engineering applications. As the structure-property relationships of plastics are discussed in the pages that follow, the reader will find it helpful to make frequent reference to this chart.

PRINCIPLES OF STRUCTURE-PROPERTY RELATIONSHIPS

Through materials science, it has been determined that thermoplastics actually contain two types of structures: *crystalline* and *amorphous.* These structures have somewhat different properties. For example, crystalline plastics tend to behave more elastically, whereas amorphous plastics flow readily and are transparent. Thermosets exhibit a third type of struc-

cept that a hydrogen atom has been replaced by a benzene ring. (This replacement would recur many times in the total macromolecule, only a small portion of which is indicated in the drawing.) Structural regularity for polystyrene is based on the benzene side group. If all the side groups appear on one side of the chain, the result is also called an *isotactic* pattern. This would still be regular, though less so than the polyethylene chain. If the pendant groups are on both sides of the chain, but still in a regular arrangement, this is called a *syndiotactic* pattern. If the side groups are randomly joined to the chain, the irregular arrangement which results is called *atactic*. In other words, atactic chains are those which do not exhibit tacticity. These three arrangements are illustrated in Fig. 4-6. The tacticity of a chain affects the molecular packing and is related to the properties of plastics.

Non-Linear Homopolymers

Structural modification can be made by causing a macromolecule to grow at several locations other than the ends. This produces a non-linear macromolecule

called a *branched polymer*. Fig. 4-7. To impart certain properties to plastics, manufacturers can actually cause the chains to adopt specific shapes such as T's, stars, or combs. Branched polymers will not pack as tightly as linear polymers. For example, so-called low density polyethylene (specific gravity = 0.92) is branched, whereas high-density polyethylene (specific gravity = 0.95) is essentially linear.

Copolymers

Another method of altering molecular architecture is to unite two different polymers within a single chain by *copolymerization*. Thus styrene and butadiene molecules can be joined to form synthetic rubber. Fig. 4-8. The result is a chain that contains alternating short segments of both molecules. The alternating sections can be joined at random, with specific pattern, or in fairly long blocks of the same polymer. Another method of copolymerization is to join chains of two different plastic monomers by graft polymerization. For example, a polybutadiene chain is grafted to a polystyrene chain to yield high-impact polystyrene. Fig. 4-9.

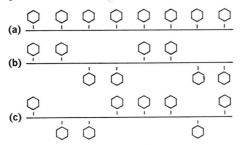

4-6. Tacticity. A segment of a polystyrene chain shows the benzene side group which forms the basis for tacticity. The side groups can be arranged in three ways. If they all lie in the same relative position, the chain is *isotactic* (a). When they form a regular arrangement, but are not all in the same orientation, the chain is *syndiotactic* (b). Side groups oriented at random result in an *atactic* chain (c).

4-7. Branched polymer. Long sections of polyethylene monomers can be attached to the carbon backbone in place of a hydrogen atom to form a branched structure rather than linear polyethylene. It is possible to produce specific shapes such as a comb, T, or star.

171

Technology of Industrial Materials

Molecular Weight

Plastics with longer polymer chains have greater molecular weights. The longer the chains, the more they tend to intertwine (as a result of the coiling described earlier). This intertwining makes the plastic more viscous, which is important in processing the plastic. In extrusion, for example, the viscosity of the plastic controls its flow and finish. Therefore manufacturers specify the average molecular weight of their material and may even produce several weight ranges of the same material.

The determination of chain size is a difficult and complex process. Detailed discussion of it is beyond the scope of this book. Basically, however, two averaging techniques are commonly employed. The *weight average* method gives the average molecular weight of the chains in a sample. If this is divided by the weight of a mer, the relative length of the chain in mer units can be determined. This value is called the *degree of polymerization.* The second method produces a *number average weight.* Because of differences in technique, the two methods produce different values and should not be interchanged. (More detailed information on molecular weight determinations, if desired, can be found in the references on page 194.)

Molecular Packing

Plastic macromolecules can be packed or grouped together into three types of structures:
- Amorphous.
- Crystalline (including semicrystalline).
- Network.

Each structure has a characteristic type of mechanical behavior. The structures and their properties will be examined.

APPLYING STRUCTURE-PROPERTY RELATIONSHIPS

Amorphous Plastics

The term *amorphous* means without a regular or definite form. When non-linear and irregular polymers solidify they have an amorphous structure. (This is not a true

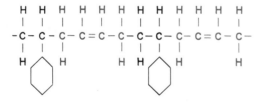

(a)   — A B A B A B A B A —

(b)   — A B A A A B B A B B B A —

(c)   — A A A A B B B B B A A A A B B B B B —

4-8. Copolymers. Styrene (in black) and butadiene (in color) can be copolymerized into the same plastic chain, forming a synthetic rubber. The chain can take one of three patterns (shown in the lines of A's and B's). The two monomers can be united in a regular pattern, forming an *alternating copolymer* (a). When no apparent pattern can be discerned, the two monomers form a *random copolymer* (b). Copolymerization can also be carried out so that fairly long sections of each chain are composed of a single monomer. This is a *block copolymer* (c).

```
A A A A A A A A A A A A A A A
              B
              B
              B
              B
              B
              B
              B
              B
              B
              B
```

4-9. Graft polymer. Polystyrene (A) can be grafted to a polybutadiene macromolecule (B). Although the chains are chemically attached, the molecular structure of each chain is preserved. Graft polymerization produces less intimate mixing of the monomers than block copolymerization.

packaging as for ketchup. By using linear chains and controlling the processing to enhance crystallinity—for example by extruding—the polyethylene that is produced will be highly crystalline and hard.

Network Structure

As mentioned earlier, a network structure forms in a plastic when the chains of the macromolecules are bonded or tied to the neighboring chains. This may occur simply through the entanglement of the chains. Such tying is dependent upon temperature and chain flexibility, and is not a chemical bond. More stable and stronger bonds are formed when the chains are chemically cross-linked. When this happens the entire structure is bonded into a single, three-dimensional network or framework. Network plastics are hard, exhibit high strength, and resist corrosion. In this text only chemically cross-linked plastics will be classified as networks, because the network is the characteristic structure of such plastics in the solid condition.

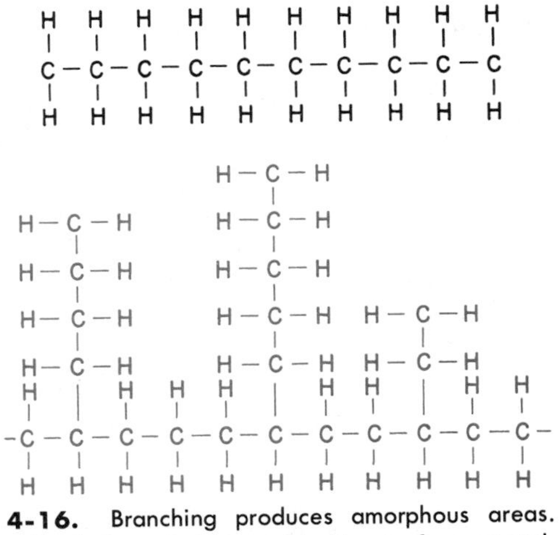

4-16. Branching produces amorphous areas. Polyethylene chains cannot align to form crystalline regions because of branched architecture. The mismatch results in amorphous regions within a normally crystalline plastic.

Cross-Linking

The oldest method of cross-linking is the vulcanization of rubber. Natural rubber is the polymer of isoprene. When bivalent sulfur atoms are introduced into the presence of isoprene chains, they displace a hydrogen atom on each of two isoprene macromolecules. Thus the chains of the isoprene are bonded into a network of hard rubber. Fig. 4-17.

Thermosetting plastics are cross-linked plastic macromolecules. A reacting organic compound can be used as the

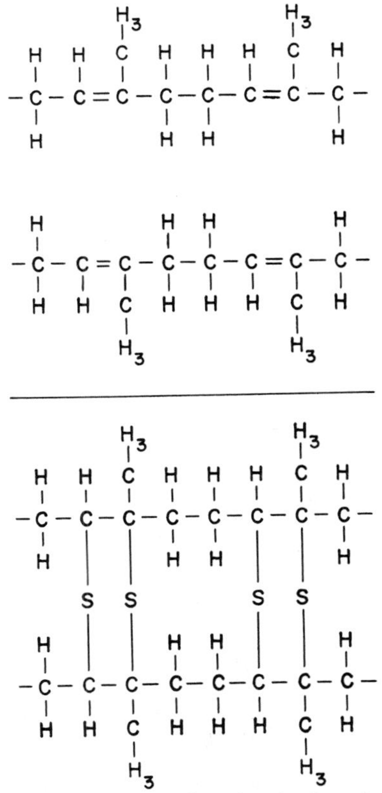

4-17. Vulcanization. The vulcanization of natural rubber occurs when chains of isoprene monomers (shown separate in upper part of drawing) are chemically linked by sulfur atoms (as shown in lower part). Note that the C=C bonds become C—C when the sulfur is present.

cross-linking agent. For example, if divinylbenzene molecules are added to polystyrene, they cross-link the chains. Fig. 4-18. This produces a network form of polystyrene. Other plastics can cross-link using oxygen. Paints, as they dry, develop a network structure through oxygen cross-linking.

Another method of cross-linking plastic is to use a monomer which is polyfunctional. Polyfunctional monomers such as epoxy can grow into a polymer chain in several directions, not merely at either end. Thus they can form a three-dimensional network as growing chains are bonded together. To initiate this type of reaction, it is necessary to add a catalyst, such as amine, to the epoxy system. The catalyst does not add to the network as the sulfur atoms do in vulcanization.

Curing

The cross-linking reaction is commonly called *curing*. After cross-linking has occurred, the plastic is difficult to form. Therefore curing is carried out after the thermoset is shaped. In industry, this is usually done by molding the plastic as a low-molecular-weight material. If a catalyst is necessary, it may be added during or immediately following molding. Then the cross-linking reaction is allowed to occur, usually with heat and pressure applied. Transfer molding, explained later, is an example of a combination of processing and curing.

Table 4-G

Effect of cross-linking frequency on the glass temperature of rubber. As the sulfur content of natural rubber increases, the frequency of cross-linking also increases. Increased cross-linking retards chain motion and thus raises the glass temperature of rubber.

Percentage of Sulfur	Glass Transition Temperature, °F
0.0	100
10.0	120
20.0	150

Mechanical Properties of Network Plastics

Network plastics are hard and strong. The properties of these thermosets result from the chemically cross-linked structure. When chains are cross-linked, they do not easily slide past each other or flow. Thus, flowing under load is minimized. In a Voigt model for thermoset mechanical behavior, the spring or elastic response would be dominant. Further, the cross-linked chains cooperate in carrying an applied load. This load-sharing capability, combined with restricted chain motion, makes thermosets hard and strong, excellent for engineering applications.

The average number of cross-links per chain has a marked effect on the behavior of network plastics. As the number of cross-links increases, the length and flexibility of chain segments between them will diminish, increasing the glass temperature and making the plastic much harder. Table 4-G. For example, natural rubber

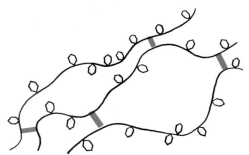

4-18. Cross-linked polystyrene. Polystyrene can be cross-linked by divinylbenzene molecules (shown in color). The polystyrene chains are bonded to form a network. Applying a load to any such chain will evoke a response from the entire network, thus raising the mechanical strength as compared to unlinked polystyrene.

Creative Packaging Company

4-28. This electric sewing-scissors kit illustrates a variety of plastic forming operations. The sturdy case (right) is made of injection molded polystyrene, with a hinged top of clear polystyrene to display the product. The unit is held in a contoured rack of thermoformed ABS (center). The complete package is suspended inside a carton by molded polystyrene foam end-caps—light in weight and high in impact resistance—for shipping. The caps are at left rear.

sheets (as for counter tops), rods, tubes, or formed shapes. Whatever the final shape, the first step in high-pressure laminating is the impregnating of the reinforcing materials with plastic resins.

In producing a flat surface, impregnated sheets, (Fig. 4-29) are stacked between two highly polished steel plates. They are then subjected to heat and high pressure in a hydraulic press which cures the plastic and presses the plies of materials into a single piece of the desired thickness. In making high-pressure tubing, resin-treated reinforcing sheets are wrapped, under tension and/or pressure, around a heated rod. The assembly is then cured in an oven. In producing formed shapes, the reinforcing material is cut into pieces that conform to the contour of the product.

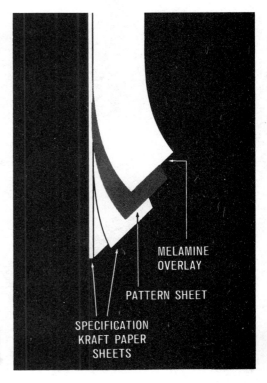

4-29. Laminating is a process used to produce tough, durable plastic sheets, such as those used on kitchen countertops. A typical structure is shown.

187

Technology of Industrial Materials

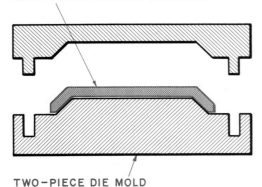

4-30. Plastic reinforcing diagram. This is one of several techniques.

4-31. One typical glass-reinforced plastic product. Automobile bodies and containers are similarly made.

British Industrial Plastics Ltd.

The pieces are fitted into the mold and cured under heat and pressure.

Reinforcing

Reinforced plastics are usually made from thermosets. They differ from high-pressure laminates in that very low or no pressure is used in the process. Reinforcing is similar to laminating in that in both methods plastic is used to bind together the cloth, paper, or glass fiber reinforcing material used for the body of the product. The reinforcing materials may be in sheet or mat form, and their selection depends on the qualities desired in the end product.

Reinforced plastics offer exceptionally high strength with low weight. Also they lend themselves to easy, economical fabrication; they require relatively little pressure and heat for curing, and they can be processed with molds and other equipment that cost far less than metal fabricating dies and machinery. Because of these advantages many products, especially large ones, are much more practical to make of plastic than of metal.

A number of techniques may be employed in the production of reinforced plastics. Fig. 4-30. In making formed shapes, impregnated reinforcing material is cut, in accordance with the shape of the finished product, into one or more pieces. The pattern or patterns are placed on a male mold in sufficient number to give the final thickness and form. The molding is completed in heated, mated dies.

As an alternative, a single mold may be used in one of the following ways:

◆ The impregnated reinforcing material may be laid up on a male mold, inserted in a rubber bag from which all air is withdrawn (so the bag presses around the layup), and cured in an oven.

Chapter 4. Polymeric Materials

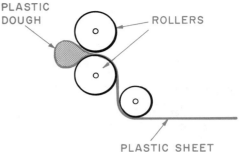

4-33. Calendering diagram.

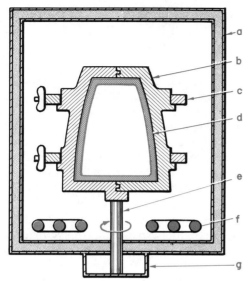

4-32. Rotational molding diagram. a. Oven b. Split mold (section). c. Clamps. d. Formed workpiece. e. Rotating shaft. f. Heating coils. g. Motor.

◆ If a female mold is used, a diaphragm is placed over the end of the mold; thus when air is withdrawn from within the mold, the diaphragm is drawn down inside the mold to press against the resin-impregnated materials.

Reinforced products are typified by GRP (glass reinforced plastics) or fiberglass items. Automobile bodies, bows and arrows, boat hulls, and storage bins are common examples. Fig. 4-31.

Rotational Molding

This technique is used to fabricate hollow, one-piece, flexible parts from vinyl plastisols or from polyethylene powders.

Essentially, rotational molding consists of placing resin into a warm mold, which is then rotated about two axes in an oven.

Centrifugal force distributes the plastic evenly throughout the mold, and the heat fuses the resin to the shape of the mold.

The finished part is extracted when the mold has cooled. This technique has several advantages, such as production of very large parts (limited only by the oven size), low mold cost, and uniform wall thickness. Representative products include balls; squeeze bulbs; hollow, flexible toys; and cisterns. Fig. 4-32.

Calendaring

Calendaring is both a method of manufacturing thermoplastic sheets and films, and applying plastic coatings to textiles and other flexible materials. The term *film* refers to thicknesses up to 10 mils, while *sheet* refers to sizes above 10 mils.

In calendaring films and sheets, plastic compound is passed between a series of three or four large, heated rollers which squeeze the material to the desired thickness. The thickness is controlled by the space between the rollers; the finish (smooth or matte) is determined by the surface of the rollers. Different materials and thicknesses require different roll arrangements, Fig. 4-33. The "F" arrangement of rollers (in which the sheet winds around several rollers arranged roughly in the shape of an F) is best suited to flexible sheets, while the "L" scheme is best for rigid sheets. Calendaring is the major method for producing vinyl and polyethylene sheets and vinyl tile.

Technology of Industrial Materials

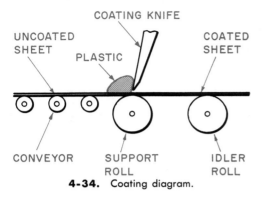

4-34. Coating diagram.

Coating

Plastic materials are frequently employed as coatings for substances such as metal, wood, paper, and leather. Methods include knife or spread coating (Fig. 4-34), spraying, calendaring, dipping, and fluidized-bed coating. In spread coating, one of the more common systems, the workpiece passes under a long blade or knife. The plastic compound is located just in front of the knife, and is evenly spread over the passing workpiece. Thicknesses are regulated by both the speed of the pass and the position of the knife.

Foaming

One of the truly unique features of plastics is the capability of producing them as *foams*. Such foams possess a broad density range (from rigid to elastic), yet they retain much of the toughness and sturdiness associated with plastic materials. *Rigid* foams are those which do not bend or sag when a reasonably large piece is held horizontally by one corner. This category includes rigid polyurethane foams used for furniture components and house construction, and also the substances commonly called *styrofoam* which are actually expandable polyethylenes and polystyrene. Fig. 4-35 is a typical foaming diagram for styrofoam. Foamed polystyrene products are commonly used as cushions for packaging fragile parts, insulation and flotation devices, and in innovative architectural applications, Fig. 4-36. Rigid polyurethanes result in denser, heavier, and tougher products which can be used in structural applications.

Flexible foam parts change shape by flexing, draping, or bending under their own weight. Examples are foamed cloth-

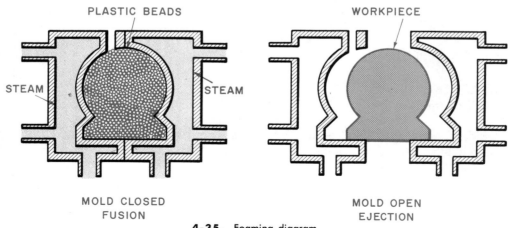

4-35. Foaming diagram.

Chapter 5

Ceramic Materials

Ceramics are essentially crystalline materials. They are a combination of one or more metals, such as aluminum, (or semimetals such as silicon) with a nonmetallic element, usually oxygen. Characteristically these substances have large molecules involving bonds in which the ionic predominates, but which are also to an extent covalent. (Review Chapter 2 for a discussion of bonding.) It is this tight bonding system which accounts for the general strength, stability, hardness, and insulation qualities of ceramics.

Not all ceramics are crystalline (glass being the prime exception) nor are they all compounds of oxygen (for example, silicon carbide). However, for purposes of illustrating the nature of ceramic materials, the discussion in this chapter will center around oxygen-base clay crystals, which are common to most ceramic compounds. Fig. 5-1. Such ceramic materials, though differing widely in their properties, have much in common:

◆ First, they are generally molded or shaped into products while they are moist and plastic.
◆ Second, after molding, these products are dried to remove the water from the clay body or mass. *Drying* increases the strength of the body, enabling it to be handled without deformation.
◆ Third, these dried articles are placed in a furnace or kiln where they are heated or fired. *Firing* removes remaining water, and a glassy bond is produced through a thermochemical action. This initial heat process is called *bisque* firing. A second *glost* firing is employed to bond an applied glaze to the workpiece.

CERAMIC STRUCTURE

The basic structural pattern of most ceramic materials is the *silicate unit* or *silicate tetrahedron* (as explained in Chapter 2.) To review briefly, in this primary unit a silicon atom is surrounded by four tetrahedrally arranged oxygen atoms. Fig.

Design Center, Prague, Czechoslovakia

5-1. These simple yet attractive salt-glazed vases are typical stoneware products. The art and science of ceramics involves the manufacture of articles ranging from fine table china and glass to sewer pipes and cement.

195

Technology of Industrial Materials

5-2. Each silicon atom has four available valence electrons, and each of the four oxygen atoms takes one. This arrangement "satisfies" the silicon but leaves the outer shells of the oxygen atoms one electron short. To compensate for this electron deficiency, the oxygen atoms link with other silicon atoms, pick up the necessary electrons, and in the process form a covalently bonded chain of silicate units. Such an arrangement appears as a fiberlike chain, much as is found in asbestos. More typically, these chains arrange themselves adjacently and are cross-linked together to form silicate sheets. These sheets are sometimes found stacked as in layered minerals such as mica, pumice, and talc.

It is precisely this sheet structure, in combination with similar sheets, which forms the basis for the clay ceramics. Fig. 5-3. The regular packing of these silicate molecules creates a matrix which readily joins with an aluminate sheet made up of aluminum and hydroxyl ions. As these two sheets mesh together, the positive aluminum ions nestle into the negative silicate ion cavities and are surrounded by six adjacent oxygen or hydroxyl ions. The resultant substance, called *kaolinite* ($Al_2O_3 \cdot 2SiO_2 \cdot 2H_2O$), is the prime ingredient in ordinary clay. Kaolinite has a strong structure in which both ionic and covalent bonding are present.

Kaolinite is a product of the decomposition of feldspar, a mineral common to rocks such as granite. Surface water, acids, and carbon dioxide percolate downward into the soil, converting the feldspar into potassium carbonate (potash), silicon dioxide (silica), and aluminum oxide (alumina). The potash is leached away, leaving the silica and alumina to combine with water to form kaolinite. This mineral frequently contains iron oxide traces which give common clay its distinctive reddish color; the presence of other impurities causes clay to assume other colors.

Kaolinite particles have the shape of tiny irregular hexagons, only .00005 of a centimetre wide and about a tenth as thick. These particles, sometimes called *platelets* or *flakes,* cling together like wet bits of paper. Slip (see Chapter 2) occurs readily in such a mass. In kaolinite, slip occurs in this way: Water molecules are attracted to the platelet surfaces by van der Waals forces, which provide a weak bond; the molecules then act as a lubricant to permit the tiny plates to slide past one another. It is this plasticity which makes it possible to mold the clay to a desired shape.

As the water is expelled through drying, the platelets interlock, and the shaped mass, though still rather weak, becomes rigid. The firing of this clay in a kiln evaporates any remaining water, and the temperature rises to the point where local melting occurs. In this melting, some of the silica combines with certain impurities to form a liquid glass "cement." When cooled, the platelets remain imbedded in a glassy medium, and together they form

● SILICON   ○ OXYGEN

5-2. The silicate unit is fundamental to clay materials. It possesses a strong ionic bond because each metal (silicon) atom gives up two electrons to each oxygen atom.

Chapter 5. Ceramic Materials

the microscopic concrete structure common to clay ceramics.

PYROMETRIC CONE EQUIVALENT

The preceding paragraph mentioned the melting of clay. The subjecting of ceramic materials to high temperatures is commonly a part of their processing. However, ceramics engineers experience some difficulty in measuring the temperature at which ceramic materials melt, because most such materials soften progressively over a range of temperatures. One method of overcoming this difficulty is to use pyrometric cones which are manufactured according to very rigid standards. When heated, these ceramic cones soften and bend before they melt, with different cones bending at different temperatures. This resistance to bending is called the *pyrometric cone equivalent* or PCE. Uses of these cones are varied, but perhaps the most common is in the firing of ceramic pottery. In such situations, the ceramist places several cones into the kiln with the material to be fired. To select the cones he consults a chart of these cone equivalents, with numbers indicating the temperatures at which the cones will slump. Some selected PCE's are shown on the chart in Fig. 5-4. When the first cone melts, the ceramist knows that the firing temperature is being approached. When the second cone melts, he knows the proper firing temperature has been reached and he begins to shut down his kiln. A higher-number cone is generally included with the other two to give some indication of possible overfiring. The use the these cones as a temperature control device is quite satisfactory because the cones react much as the product being fired does. Modern industrial practices include the use of pyrometers as well as cones in firing operations.

RAW MATERIALS

Although the various clays which are used in ceramics vary considerably in both composition and color, they do possess some common properties:

- When moist they are plastic and moldable.
- When dry they become rigid.
- When fired they become strong, hard, and permanently nonmoldable.

The following are some of the more common raw materials used in the production of clay bodies:

- *Kaolin.* A fine, white clay, nearly pure kaolinite, which is the basis for china and porcelain. Also known as *china clay.*

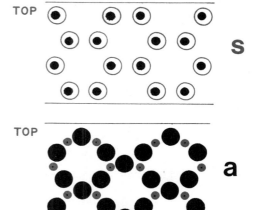

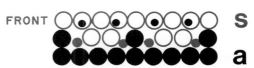

5-3. The silicate sheet (s) and the aluminate sheet (a) reveal a basic tetrahedral pattern of molecules in their top views. When placed face to face, these two sheets mesh to form an ionically bonded laminate of kaolinite, as shown in the front view. Stacked kaolin sheets are held together with weaker hydrogen and van der Waals bonds. This diagram has been simplified for clarity.

Technology of Industrial Materials

Cone	°F
09	1710
01	2090
1	2120
6	2250
8	2300
10	2380
11	2420
12	2440
13	2460
20	2790
30	3000
35	3250

5-4. Selected pyrometric cone equivalents. Typical cone arrangements are shown in the photograph, indicating cone appearance before (right) and after firing (left).

- *Grog.* Prefired clay. Gives a porous texture to materials such as refractories, thus preventing the extension of cracks and fissures.
- *Flint Clay.* A hard, low-plasticity type of fireclay often found in coal deposits. It is a valuable clay additive because of its low firing shrinkage.
- *Feldspar.* An alkali-aluminosilicate present in almost all igneous rocks. Feldspar is not a clay but a rock widely used as a fluxing ingredient in clay mixes.
- *Fireclay.* Any naturally occurring refractory clay, normally consisting of kaolinite, free silica, and a few impurities. It is a very plastic clay and is refractory because of its low flux content.
- *Ball Clay.* A fine, dark, sedimentary clay of excellent plasticity and dry strength. It is widely used in making whiteware to increase plasticity and strength. Such clay reduces translucency in a ceramic, although it fires to a white or cream color.
- *Glacial Clay.* A dense, gritty, red-firing clay derived from rock particles washed from glacial deposits. Used extensively for brick and other structural ceramics. Also called *brick clay.*
- *Shale.* A hard, nonplastic sedimentary clay with the property of low shrinkage.
- *Slip Clay.* An impure clay of glacial origin, containing enough flux to melt at moderate temperatures. Used as glaze on ceramic bodies.

These clay minerals are seldom used individually in the makup of ceramic bodies. Instead, such bodies contain a mixture of different clays and nonclay additives (such as grog or feldspar) in order to achieve a product possessing the desired qualities. For example, ball clays are added to provide the required plasticity for forming, and to give additional strength to green and dried clays. Kaolins enable the body to be fired with a minimum of cracking and warping. Feldspar is a principal *vitrifying* (glass-forming) and fluxing ingredient. Flint acts as a nonplastic filler in green clay and as a refractory in the fired body. Flint also acts to decrease shrinking and warping during drying and firing cycles. Such are the

vides a more workable plastic body as well as a stronger dry body.

A common spark plug is a classic example of such a refractory insulator. Fig. 5-10. The most important function of a spark plug insulator is to prevent the high-voltage ignition current from flowing in any direction except acrosss the spark gap between the electrode wires. In addition, the insulator must be mechanically strong to withstand normal handling and engine vibration.

The voltage requirements of electrical transmission equipment have risen from 150 kV in 1913 to over 550 kV today. Consequently, the sizes and specifications for these huge insulators have risen correspondingly. High-quality ceramics make possible the transmission of high voltages.

Glass

The great bulk of common glasses are obtained by mixing silica with other minerals and melting them at high temperatures. The resultant hot liquid cools to a hard, rigid mass without crystallizing—that is, the atoms never arrange themselves into an orderly crystalline lattice. Fig. 5-11. In the hardened glass, this atomic pattern resembles the arrangement in a glass liquid state, and for this reason glass is frequently referred to as a hard liquid. Glass remains in this noncrystalline state no matter how hard and rigid it becomes when cool.

The large number of differences in glass is produced by employing different ingredients in the batch mixes, and by varying the manufacturing processes. Thus there are glasses hard as precious stones and others as soft as down; some as strong as steel and others as fragile as eggshells.

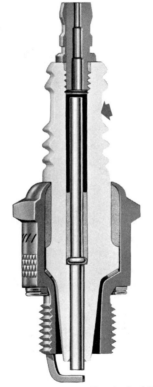

Champion Spark Plug Company

5-10. A spark plug insulator (arrow) must prevent leakage of electrical currents ranging from 10,000 to 30,000 volts, at temperatures varying from −50 to +1700° F.

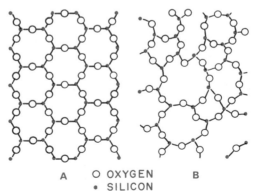

A ○ OXYGEN B
 • SILICON

5-11. Crystalline (a) and noncrystalline (b) forms of silica. Glass atoms never achieve the orderly arrangement of a crystal.

Technology of Industrial Materials

Corning Glass Works

5-12. The spectrum of glass products runs from window glass to television picture tubes, and from electronic circuitry to laboratory apparatus.

From these are made a broad variety of glass products. Fig. 5-12. However, commercial glasses can be classified into six basic types according to ingredients and special properties. Table 5-A. (Selected physical properties are given in Appendix E.)

Special Glasses

Differences among the six basic types of glass, as explained, are produced by varying the ingredients and the manufacture. By still further variation it is possible to produce types of glass with truly remarkable properties, far superior to the transparency, hardness, and insulation qualities of the six basic types. These unusual glasses are therefore not classed as basic types but are categorized as special glasses. Figs. 5-13 and 5-14. In *coated glasses*, a thin metallic oxide coating is fused to the surface of the glass. Because this film can conduct electricity, it can be used for domestic heating and lighting panels and electronic resistors. *Glass fibers* can be drawn from several of the basic compositions. The strength and insulation properties of this product offer many applications in thermal and acoustical insulation. It can also serve as a tough matrix when mixed with plastic resins to form the familiar glass-reinforced plastics. *Glass foams* are made by heating a mixture of pulverized glass and a foaming agent. The resultant rigid foam mass is almost as light as cork, can be cut with woodworking tools, and makes an excellent insulator. *Glass ceramics* (marketed as Pyroceram) are crystalline materials made from noncrystalline glass. Their outstanding thermal shock resistance makes them desirable as materials for cookware, missile

Chapter 5. Ceramic Materials

Table 5-A
Basic glass chart.

Glass Type	Typical Composition	Characteristics	Applications
1. Soda-lime-silica	SiO_2 (silica) 72% Na_2O (soda) 15% CaO (lime) 9% MgO (magnesia) 3% Al_2O_3 (alumina) 1%	Most common of all glasses, accounting for nearly 90% of glass used in the world. Easy to melt and shape; inexpensive; poor thermal and chemical resistance.	Window glass, plate glass, containers, building blocks, light bulbs.
2. Borosilicate	SiO_2 (silica) 80% B_2O_3 (boric oxide) 14% Na_2O (soda) 4% Al_2O_3 (alumina) 2%	Good resistance to thermal shock and corrosion, low thermal expansion. Hard to work because of high softening temperature.	Laboratory apparatus, piping, sealed beam headlamps, telescope mirror blanks, ovenware.
3. Lead-alkali-silicate	SiO_2 (silica) 68% PbO (lead oxide) 15% Na_2O (soda) 10% K_2O (potash) 6% CaO (lime) 1%	High index of refraction, good infrared transmission and electrical properties.	Crystal glass for tableware; thermometer and neon sign tubes; optical lenses.
4. Fused silica	SiO_2 (silica) 99% +	Simplest glass chemically; made directly from silica. Exceptional radiation resistance, optical qualities, and thermal shock resistance. Low thermal expansion, at a rate of 5 as compared to 90 for soda-lime-silica glass.	Lightweight, space-borne telescope mirrors; space vehicle windows; laser beam reflectors.
5. 96% silica	SiO_2 (silica) 96% B_2O_3 (boric oxide) 3% Other oxides 1%	Made by removing the nonsilicate ingredients from borosilicate glass. Expensive. Exceptional thermal properties and chemical resistance (can be heated cherry red and plunged into ice water without damage.)	Missile nose cones, space vehicle windows, optical lenses, chemical glassware.
6. Aluminosilicate	SiO_2 (silica) 60% Al_2O_3 (alumina) 15% CaO (lime) 10% MgO (magnesia) 7% B_2O_3 (boric oxide) 8%	Outstanding thermal shock resistance, good electrical and chemical properties, very high softening temperatures.	High temperature thermometers, stove top cookware, laboratory apparatus.

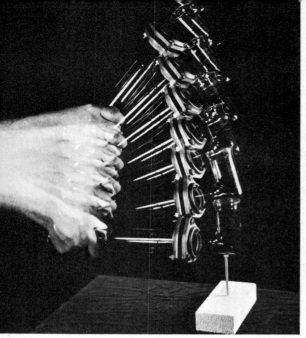

Corning Glass Works

5-13. A section of specially tempered glass pipe is strong enough to hammer a nail into a block of wood without shattering.

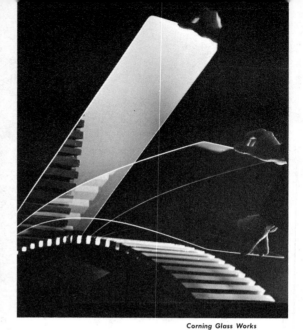

Corning Glass Works

5-14. Chemically strengthened glass exhibits a high flexural strength, permitting the fabrication of lightweight articles without a sacrifice in strength.

nose cones, and burner covering surfaces for electric stoves. Fig. 5-15.

Porcelain Enamels

Porcelain enamels are inorganic glass coatings for metals, applied by fusing powdered glass into a continuous layer on the surface of the metal. They are similar to glazes in that they are basically a silicate glass coating, but they are normally fired for shorter periods and at lower temperatures. Such enamels are desirable materials for coating metals because of their ability to resist acids and water as well as their excellent resistance to abrasion and heat. In addition, they have desirable visual properties that include a full range of color, finish, texture, and opacity.

While enamels are basically glass materials, flux modifiers must be added to lower their melting point so as not to interfere or coincide with the melting point of the metal which is being coated. Typical modifiers are boron, fluorine, and lithium.

Lead was at one time a popular additive which produced an enamel of brilliant color; however, today for reasons of health its use in enamels is quite limited. Pigments and opacifiers are added to produce the desired colors.

Many metals can be coated with enamel materials. Gold, silver, copper, and copper alloys are popular materials in the production of enameled jewelry. Fig. 5-16 gives the three basic methods employed to produce a variety of color patterns, thus enhancing the beauty and interest of the jewelry.

The familiar enamel coatings on electric appliance housings, such as refrigerators, electric stoves, and washing machines, are produced by the sheet-steel enameling process. Fig. 5-17. In this process sheet steel is pressed into the desired shape and thoroughly cleaned or pickled by dipping into a bath of dilute acid. After rinsing and drying, a ground coat of enamel is applied, either as a powder or as a *liquid*

Chapter 5. Ceramic Materials

Portland Cement Association

5-22. Pre-cast concrete modular units are made at the plant and hauled to the building site for assembly.

and plastering. These materials lack the strength and durability of portland cement mixtures but are lighter and easier to use for the special applications cited above. Most of them are based upon *gypsum*, a stone comprised of calcium sulfate and water.

- *Plaster of paris* is made by heating gypsum to dehydrate it, leaving a white powder. When water is added to the powder, it hardens to a stonelike mass similar to the original gypsum. It is used for investment moldmaking and small statuary castings.
- *Keene's (Keenan's) cement* is of a composition similar to plaster of paris except that alum is added as an accelerator. Stronger and more water resistant, it is valuable as a grout for tile joints and for plastering.
- *Gypsum plasters* are used as thin protective wall coatings, troweled or sprayed on a prepared surface. They are basically plaster of paris mixes with sand, lime, and horsehair added for strength and ease of application. This material is also used in the production of plaster wall boards.

CERAMIC MANUFACTURING PROCESSES

The processing of ceramic materials into usable products is simplified by their natural plasticity. Furthermore, it is obviously easier and more economical to mold these materials while they remain unfired, for after firing further processing becomes most difficult. Ceramics are worked in a fashion similar to metals and plastics; many of the processes are therefore familiar. Clays and concretes are, in many respects, worked differently from glass in that the latter is generally molded while hot.

The following charts and illustrations identify and describe some of the more important techniques used in processing clay and concrete materials. Glasses are treated separately in subsequent paragraphs.

CUTTING—Removing or separating pieces of material from a base material.

Type	Description	Processes or Products
Shearing	Separating a moist ceramic workpiece by forcing a single cutting edge through it.	Wire slicing workpieces to size. Trimming finished pieces with knives or other cutters.
Turning	Shaping a dry or semidry workpiece by revolving it against a fixed cutting tool; a material removal process. Fig. 5-23.	Turning electrical insulators on vertical or horizontal lathes. Cutting a "foot" on a piece of ceramic artware.
Abrading	Separating or removing material through the action of mineral particles against a workpiece.	Grinding, sanding, and polishing articles of clay and concrete.
Sawing	Separating a workpiece by forcing it against a rotating, circular blade with hardened teeth.	Sawing brick and concrete products to size.

FORMING—Shaping a material without adding or removing any of the material.

Type	Description	Processes or Products
Extrusion	Forcing plastic ceramics through a forming die to a desired cross-sectional shape.	Brick, hollow tile, tubes, and rods.
Casting	Pouring liquid ceramic into a cavity and allowing it to set.	Small ceramic parts; concrete parts, benches, walks, light standards, building panels.
Slip-casting	Pouring clay slurry (slip) into an absorbent mold; a deposit forms inside the mold, excess slip is poured off. Fig. 5-24.	Sinks, toilets, and whitewares of irregular shape.
Jiggering	Placing a moist clay slab on a revolving plaster mold, and forcing it against the mold with a profiling tool. Fig. 5-25.	Symmetrical articles such as plates, saucers, and bowls.
Pressing	Forcing moist clay (wet pressing) or dampened ceramic powders (semidry pressing) into simple or sectional molds.	Wall tile, electrical insulators, and electronic parts.
Hand forming	Placing moist clay on a revolving potter's wheel and pressing the clay into shape with the hands (throwing); or hand-building shapes from coils or slabs.	Ceramic artware, special laboratory apparatus.

FASTENING—Joining materials together permanently or semipermanently.

Type	Description	Processes or Products
Adhesion	Permanently joining like or unlike materials with glue or cement.	Glue and epoxy cement repairs, grouting tile, mortaring bricks.
Cohesion	Permanently joining by mixing like materials together and fusing them into a homogeneous, continuous mass. Fig. 5-26.	Clay welding (mixing at a joint), slip cementing, firing clay products.

FINISHING—Treating a material to improve its appearance and/or protect its surface.

Type	Description	Processes or Products
Decorating	Embellishing a surface. Fig. 5-27.	Applying stain, enamel, decals.
Glazing	Coating objects with clay slip to produce a decorative or protective glassy surface. Fig. 5-28.	Applying glazes by brushing or dipping.

5-23. These lightning arrester housings, weighing 1800 pounds, are being turned on special vertical lathes. As the dried blank rotates, the cutting tool (not visible) starts at the top and moves downward to produce the rotary grooves. These grooves, called petticoats, serve to increase the surface distance an electric charge must travel to create a short circuit.

Ohio Brass Company

5-24. A pitcher made by slip casting. After the deposit is formed and the excess slip is poured off, the workpiece shrinks and can be easily removed.

Wedgwood

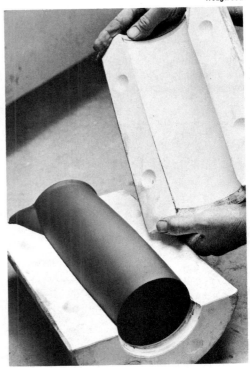

Syracuse China

5-25. Forming a plate by the jiggering process. Note the profiling tool (arrow) which controls the external shape. Jollying is a similar process used to form the internal shapes of cups and deep bowls.

Wedgwood

5-26. Cohesion-fastening a handle to a teacup by sticking them together with clay slip. After firing, the handle and cup are joined permanently.

5-27. Printing a decoration on a dinner plate with a parabola-shaped gelatin pad. Hand painting and applying decals are other decorating techniques.

Wedgwood

5-28. Dip-glazing a dinner plate. Glazes can also be painted or sprayed on a surface. Glazing is followed by glost or glaze firing to fuse the glaze to the workpiece.

Wedgwood

Chapter 6. Wood

secondary wall, S_1, the network becomes more precise. By the time the S_2 layer of the secondary wall is reached, the lamellae run almost parallel to each other in a spiral around the cells. This (S_2) layer is the thickest of the several in the cell wall and has the greatest effect on how the cell behaves. The smaller the angle which the fibrils make with the long direction of the cell, the stronger is the cell. Finally, in the innermost layer of the cell wall, S_3, the lamellae are once again in a netlike arrangement.

These several adjacent layers are bonded by the gluelike substance *lignin*. While the cellulose gives toughness to the cell wall, it is the lignin which gives it rigidity. Lignin is not a fibrous composition; however, interspersed among the lamellae and surrounding them, it creates a wall similar in structure to reinforced concrete. Fig. 6-3. Thus various kinds of cells are formed:

- Parenchyma—cells for the storage of starch and sugar.
- Tracheids—long, pointed supporting cells.
- Vessels—wide, nutrient-conducting cells.

These cells develop and arrange themselves into two distinct structural patterns—the hardwoods, or the broad-leaved trees, and the softwoods, or the cone-bearing trees, which have needles or scalelike leaves. Typical hardwoods, such as the oaks, maples, and birches, characteristically drop their leaves in the autumn. Softwoods, such as pines, firs, and spruces, are evergreen.

Porosity

The separation into these two groups does not mean that all hardwoods have harder wood than the softwoods. The true differences lie in cellular structure. Figs. 6-4 and 6-5. An examination of these two illustrations reveals a number of interesting differences and similarities. For example, there is a greater variety of cells in the hardwood, and these cells are not arranged in the orderly, radial rows characteristic of the softwood. Another important difference is that hardwoods have vessel segments; softwoods do not. When viewed in cross section, these vessels are called pores. Hardwoods are therefore called porous woods; softwoods are said to be nonporous in that they do not contain vessels. Furthermore, in some hard-

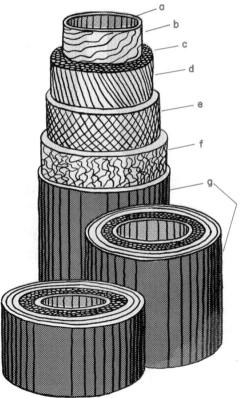

6-3. Section of wood fiber construction. a. Cell cavity (lumen). b. S_3 layer. c. Lamellae. d. S_2 layer. e. S_1 layer. f. Primary wall. g. Lignin intercellular layer (true middle lamella).

Technology of Industrial Materials

woods—such as birch and yellow poplar—the vessels are of the same diameter in both the earlywood and latewood. Such woods are called *diffuse-porous,* because the vessels are scattered throughout the entire growth ring. In other hardwoods—chestnut, oak, and ash, for example—the vessels are narrower in the latewood than the earlywood. Since these differences in vessels appear as rings in a log cross section, such woods are called *ring-porous.* Ring-porous woods usually require a wood filler in a finishing operation.

Vessel segments develop from single cells. These segments are joined end-to-end to form the vessels which serve as the main passageway for liquids moving from the roots to the crown (leaves and branches) of a tree. In softwoods, resin ducts are formed when a space among several cells expands to form an enlarged opening in the wood. Sticky resin is released from the cells lining the duct.

In summary, it is the structure, variety, and organizational pattern of the wood cells which give hardwoods and softwoods their characteristic appearances and properties.

PROPERTIES OF WOOD

To provide a basis for selecting a suitable wood for a specific design application—this is the chief reason for studying wood properties. What factors bear upon the decision to employ walnut in the construction of a fine chair, birch for cabinet doors, fir or pine for house construction, or spruce for pulpwood? In all such situations, the decision is based upon a knowledge of the properties of the wood material, in comparison with the specifications for a product design. Certain

U.S. Forest Products Laboratory

6-4. Cell structure of minute block of softwood, showing the different kinds of cells and other features. a. Cross-sectional face. b. Radial face. c. Tangential face. d. Annual ring. e. Earlywood. f. Latewood. g. Wood ray. h. Vertical resin duct. i. Horizontal resin duct.

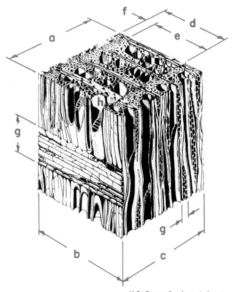

U.S. Forest Products Laboratory

6-5. Wood cell structure of hardwood. a. Cross-sectional face. b. Radial face. c. Tangential face. d. Annual ring. e. Earlywood. f. Latewood. g. Wood ray. h. Vessel.

Chapter 6. Wood

woods have outstanding physical appearance, while others are tough, rot-resistant, or aromatic. The following pages present a study of wood properties, with particular attention to their importance in product applications. A good beginning will be to examine the way moisture content and specific gravity affect the other properties and characteristics of wood.

Moisture Content

Wood materials have an affinity for water in both vapor and liquid forms, absorbing and losing moisture according to variations in temperature and humidity. *Moisture content* is the term which describes the total amount of water in a given piece of wood. This measurement is expressed as a percentage of the ovendry weight of the wood. These values are determined according to the following formula:

$$\text{MC} = \frac{\text{weight of water in wood}}{\text{ovendry weight of wood}} \times 100$$

Measures of moisture content are important because of the effects of moisture on both the mechanical and nonmechanical properties of wood materials.

Of special importance is the fact that wood, like many other materials, shrinks as it loses moisture and swells as moisture is absorbed. As wood dries, it shrinks most in the direction of the annual rings (tangentially), somewhat less across these rings (radially), and very little along the grain (longitudinally). The combined effects of tangential and radial shrinkage are shown in Fig. 6-6. Wood which contains grain irregularities will frequently shrink in a distorted fashion longitudinally, resulting in warped or bowed boards. It will be noted that quartersawed lumber has the most desirable shrinkage pattern.

Specific Gravity

This measure is the ratio of the weight of a given volume of wood to that of an equal volume of water at a standard temperature. Because the weight of wood varies with changes in moisture content (due to shrinking and swelling), *specific gravity* has to be measured under specified

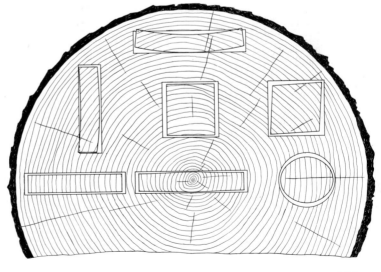

U.S. Forest Products Laboratory
6-6. Characteristic shrinkage and distortion of flats, squares, and rounds as affected by the direction of the annual rings. Tangential shrinkage is about twice as great as radial.

221

Technology of Industrial Materials

conditions. This measurement is obtained from the formula:

$$G \text{ (specific gravity)} = \frac{\text{ovendry weight of wood}}{\text{weight of displaced volume of water}}$$

This value varies for different woods according to the amount of cell wall material and the sizes of the cells.

The term *density* must be distinguished from specific gravity. Density refers to the *weight per unit volume,* as expressed in pounds per cubic foot. It is found by the equation:

$$\text{Density} = \frac{\text{weight of wood with moisture}}{\text{volume of wood with moisture}}$$

Both specific gravity and density have influence on the strength properties of wood and the suitability of various woods for specific application.

Physical Properties

As described earlier, wood consists primarily of relatively long, hollow cells running lengthwise in the trunk and branches. These cells, as well as those which radiate from the central axis, are cemented together by lignin. This structure, resembling a bundle of hollow tubes cemented together, accounts for many of the unique physical characteristics of wood materials.

Cell structure, besides influencing the mechanical properties, also affects the appearance of wood. The varied arrangement in cellular structure which is visible on the cross-sectional face of a log is what gives the wood its grain and texture. These terms are often used interchangeably, but *grain* actually refers to both the annual rings and the direction of fibers longitudinally—that is, straight or spiral grain. The terms *open* and *closed* grain indicate the relative size of the pores. *Texture* refers more specifically to the finer structure of wood.

The grain pattern, which greatly affects the decorative features of wood materials, is itself influenced by the method in which the log is sawed into boards. *Quartersawed* and *plainsawed* boards are shown in Fig. 6-7. Note the unique grain pattern for each. In quartersawed (or slash-cut) lumber, the annual rings intersect the surface at angles between 45° and 90°. In plainsawed pieces, the rings intersect at angles less than 45°. Fig. 6-8 illustrates how cut affects warping in lumber.

The color of wood and possibly the presence of knots combine with the grain structure to provide the familiar beauty of wood. Perhaps the most fascinating visual aspect of wood is the tremendous range of possible patterns.

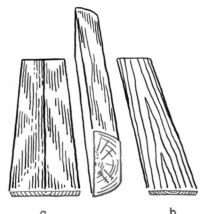

U.S. Forest Products Laboratory

6-7. Grain features of (a) quartersawed, and (b) plainsawed lumber. Note that the type of cut affects the kind and degree of shrinkage in a board.

Thermal Properties

Thermal conductivity of wood is a measure of the amount of heat, expressed in British thermal units, that flows in one

free of such defects; however, the presence of defects does not necessarily prevent lumber from being acceptable for many uses.

Most hardwood boards are not used in their entirety, but are cut into smaller pieces suitable for furniture and other products. A description of the standard hardwood grades for this kind of lumber is given in Table 6-A. The highest grade is termed *Firsts,* and the next grade *Seconds.* First and Seconds, (general written "FAS"), are in practice combined in one grade. The third grade is termed *Selects,* followed by *No. 1 Common, No. 2 Common, No. 3 A Common,* and *No. 3 B Common.* It should be noted that hardwood flooring is graded under separate standards, according to species.

Softwood lumber is graded according to the rules of a number of different associations. In order to eliminate grading differences amoung these several associations, softwood grades are based upon a set of simplified rules known as the American Lumber Standards. These standards in-

6-9. Typical cutting scheme for a Douglas fir log 42" in diameter.

Table 6-A

Standard hardwood grades. (Inspection to be made on the poorer side of the piece, except in Selects.)

Grade and lengths allowed (feet)	Widths allowed	Surface measure of pieces	Amount of each piece that must work into clear-face cuttings	Maximum cuttings allowed	Minimum size of cuttings required
Firsts:[1] 8 to 16 (will admit 30 percent of 8- to 11-foot, 1/2 of which may be 8- and 9-foot).	Inches 6+	Square feet 4 to 9 10 to 14 15+	Percent 91 2/3 91 2/3 91 2/3	Number 1 2 3	4 inches 5 feet, or 3 inches by 7 feet.
Seconds:[1] 8 to 16 (will admit 30 percent of 8- to 11-foot, 1/2 of which may be 8- and 9-foot).	6+	4 and 5 6 and 7 6 and 7 8 to 11 8 to 11 12 to 15 12 to 15 16+	83 1/3 83 1/3 91 2/3 83 1/3 91 2/3 83 1/3 91 2/3 83 1/3	1 1 2 2 3 3 4 4	Do.
Selects: 6 to 16 (will admit 30 percent of 6- to 11-foot, 1/6 of which may be 6- and 7-foot).	4+	2 and 3 4+	91 2/3 (²)	1	Do.
No. 1 Common: 4 to 16 (will admit 10 percent of 4- to 7-foot, 1/2 of which may be 4- and 5-foot).	3+	1 2 3 and 4 3 and 4 5 to 7 5 to 7 8 to 10 11 to 13 14+	100 75 66 2/3 75 66 2/3 75 66 2/3 66 2/3 66 2/3	0 1 1 2 2 3 3 4 5	4 inches by 2 feet, or 3 inches by 3 feet.
No. 2 Common: 4 to 16 (will admit 30 percent of 4- to 7-foot, 1/3 of which may be 4- and 5-foot).	3+	1 2 and 3 2 and 3 4 and 5 4 and 5 6 and 7 6 and 7 8 and 9 10 and 11 12 and 13 14+	66 2/3 50 66 2/3 50 66 2/3 50 66 2/3 50 50 50 50	1 1 2 2 3 3 4 4 5 6 7	3 inches by 2 feet.

Chapter 6. Wood

6-19. Typical wood-frame house construction.

products are examined to illustrate the suitability of a wood material to a given task.

Structural Applications

Home construction provides instructive examples of wood materials selection. Figs. 6-18 and 6-19. Studs, joists, planks, and sheathing are selected for strength and ease of on-site working. In the interior, doors, wood frames, cupboards and cabinets, and flooring all suggest the utilization of wood for both beauty and function. Fig. 6-20 shows a handsome application of architectural grillework. Heavy

Technology of Industrial Materials

6-20. Architectural interior uses of woods include room dividers and panels similar to this striking example.

6-21. Many case-constructed pieces of furniture are made from teak plywood panels for dimensional stability.

timbers are employed in building foundations and as forms for poured concrete structures.

Furniture

The beautiful grains and colors of wood make it an especially desirable material for furniture construction and cabinetwork. Wood has for centuries enjoyed a prominent position in this regard because of its workability and adaptability to a variety of settings.

Fig. 6-21 illustrates a classic application of wood in furniture making. This storage case of hand-rubbed teak is a functional unit, handsome in its simplicity and warmth. The rectangular lines are especially suited to efficient mass-production cutting techniques. The formed plywood chair, Fig. 6-22, though less easy to mass-produce, illustrates the variability of the material. In its production a plywood sheet is formed in two directions, resulting in a strong, attractive, and comfortable

6-22. The seat and back of this chair are made from formed plywood panels.

Chapter 7. Miscellaneous Materials

7-1. In machine spinning, the fibers are twisted to form firmly bound yarn. This combing machine removes short fibers, leaving only the longest and strongest for linen yarn.

American Textile Manufacturers Institute, Inc.

turing processes employed, spinning, weaving, and knitting are the most important.

Spinning

Fibers, to be usable in textile manufacture, must be rendered into long strands called *yarns*. Such yarns are made by carding and combing the fibers to separate them, provide a parallel orientation, and remove short unmanageable fibers. During subsequent processes, the fibers are drawn together, and sufficient twist is introduced to bind the yarn firmly together. Fig. 7-1. There are several spinning methods, each employed according to the type of fiber being worked. For example, *flyer spinning* is used for linen, *ring spinning* for cotton, and *mule spinning* for wool. After spinning, the yarns are dyed and wound on bobbins, ready for weaving, knitting or for use as sewing threads.

Weaving

Weaving is the process of making fabrics from two sets of threads which interlace at right angles. Fig. 7-2. Those running the length of the cloth fabric are called the *warp*, and those lying across the width of the piece are called the *weft*. There are numerous weave patterns, such as hopsack, herringbone, tweed, and satin. These weave systems produce fabrics which are loose or close, soft or hard; they also provide a broad range of interesting patterns, colors, and textures. For example, canvas or duck is a tightly knit, firm cloth made from cotton or linen. This firmly woven, strong material is ideal for sailcloth and tents. Pile fabrics, such as velvet, have threads projecting from the ground (basic surface) of the fabric and provide a smooth, luxurious texture. Weaving is an old art, but the basic warp and weft systems remains unchanged. Modern technology provides the means for producing a broad spectrum of woven fabrics.

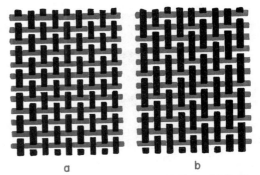

7-2. Typical weave patterns: (a) Plain. (b) Twill.

Technology of Industrial Materials

Knitting

Knitted fabrics are especially desirable in applications requiring flexibility and porosity, such as in underwear, sweaters, and stockings. In knitting, the loop is the basic structural unit; the loops are organized in rows running both the length and width of the fabric piece. Fig. 7-3. Hand knitting is still a popular art, but modern machines are used for commercial production. Fig. 7-4.

Other textile manufacturing processes include decorative *lace making; felting,* which produces an extremely tight and impermeable fabric for clothing and industrial uses; and *bonding,* in which nonwoven fabrics made directly from fibers are fastened to a backing.

Finishing Manufactured Textiles

Fabrics must generally undergo finishing treatments before they can be used. Such treatments include *washing, dyeing, printing,* and *drying.* However, other finishing processes are performed in certain situations:

- *Mercerization* imparts a permanently lustrous finish to fabrics such as cotton and linen by a treatment with caustic soda.
- *Singeing* removes fluff and loose fibers by passing the fabric over gas flames.
- *Waterproofing* involves applications of silicone, rubber, waxes, and other chemicals to produce rain-resistant or rain-proof materials.
- *Sanforizing* machines steam the fabrics and then rapidly stretch and shrink them on a series of convex and concave rollers. Cottons and linens are sanforized to prevent stretching or shrinking as they are washed.
- *Bonded* fabrics, mentioned above, are in fact two separate textiles joined permanently back to back—for example, a soft, absorbent material used to line coarse, waterproof rainwear.
- *Laminating* is bonding textiles to plastic foam liners.

LEATHER

From an industrial point of view, leather is classified as a *flexible sheet* material, similar to fabrics, paper, or sheet rubber and plastic. This flexible sheet of material can be of natural origin (natural leather) or synthetic (man-made leather). The raw material for making natural leather is the skin of certain animals, while certain plastics and rubbers are used for manufacturing synthetic leathers.

Leather

In the leather industry the pelts of large animals such as cows, steers, horses, and buffalos are called *hides*; the pelts of small animals such as calves, sheep, goats, and pigs are called *skins.* For the purpose of this text the term *hide* will be used for all pelts. About 80% of the hide consists of complex nitrogenous organic compounds called *proteins.* A certain protein called *collagen* has molecules which are made up

7-3. Typical loop knitting pattern. Knitting may be done with single or multiple threads.

7-9. Interior view of a modern shoe plant.

100m²/hr). Each type of leather has its own PV; for example, leathers used for shoe uppers must have a PV between 2.000–18.000 grams/100m²/hr.

The ability to mold leather and shape it on a last renders it useful for many applications including shoes, gloves, and other garments. Even though the individual proteinous fibers are relatively weak, leather exhibits good tensile and tear strength. This is due primarily to the arrangement of the fibers into bundles which permit progressive distribution of stresses over all fibers. Leather is flexible and can be flexed several million times before it cracks. This property provides leather with the characteristics required in such uses as shoes, upholstery, and transmission belts.

Synthetic Leathers

Many types of synthetic (man-made) leathers are produced today as substitutes or supplements to natural leather. However, although synthetic leathers having superior individual properties (such as tensile strength) are produced, no synthetic leather has been developed with the combination of properties of natural leather. All synthetic leathers are composite materials made by a combination of fabrics,

Table 7-B
Types of leather and typical applications.

Skin Origin	End Use Application
Cow and Steer	Shoe & boot uppers, soles, insoles, linings; patent leather; garments; work gloves; waist belts; luggage and cases; upholstery; transmission belting; sporting goods; packings
Calf	Shoe uppers; slippers; handbags & billfolds; hat sweatbands; bookbindings
Sheep and Lamb	Grain & suede garments; shoe linings; slippers; dress & work gloves; hat sweatbands; bookbindings; novelties
Goat and Kid	Shoe uppers, linings; dress gloves; garments; handbags
Pig	Shoe suede uppers; dress & work gloves; billfolds; fancy leather goods
Deer	Dress gloves; moccasins; garments
Horse	Shoe uppers; straps; sporting goods
Reptile	Shoe uppers; handbags; fancy leather goods

New England Tanners Club

Technology of Industrial Materials

E. I. duPont de Nemours and Company

7-10a. The structure of Corfam®, a synthetic leather, varies from relatively loose on the underside (at the bottom of this cross-section view) to much tighter at the surface. Such a structure can be described as *gradient*. Corfam® transmits moisture vapor in much the same way as natural leather.

plastics, paper, and rubber. The chemical and mechanical processing of synthetic leather is extremely complex; companies in this business usually keep the details of their processes secret.

The closest competitor or supplement to natural leather is the synthetic *Corfam®*, developed by duPont. In the production of Corfam®, as in the production of most synthetic leathers, the fiber-forming materials used are polyester, polyamide, and

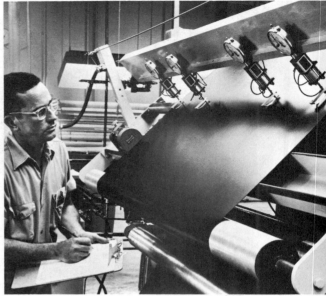

E. I. duPont de Nemours and Company

7-10b. Quality inspector of finished Corfam® sheet checking uniformity of texture, color, and thickness.

acrylic polymers. The production of most synthetic leathers involves the following steps:

- Preparation of the polymeric fibers.
- Formation of a mat with the polymeric fibers.
- Mechanical and chemical treatment of the mat.
- Application of binders and other ingredients to the mat.
- Consolidation and coating.
- Introduction of porosity (permeability).
- Modification of the surface for providing its texture.

The resulting product is a synthetic leather with many of the properties of natural leather. Fig. 7-10.

COMPOSITE MATERIALS

When two or more materials are combined and retain their identities by not

Chapter 7. Miscellaneous Materials

dissolving or merging completely into each other, the result is a *composite material*. Such a material has different properties and composition than any of the materials which were originally combined.

The materials combined to form a composite are called the *constituents* of the composite. Since the constituents retain their individual identities, the properties of the composite are influenced by the properties of the constituents. Fig. 7-11. Thus, a composite can have the strength of one constituent, the light weight of another, and the chemical, thermal, and electrical properties of a third. Numerous property combinations are possible.

Composite materials may be of natural origin (natural composites) or they may be man-made. Examples of natural composite materials are wood, leather, bones, and certain stone formations. Glass and cement are two of the many man-made composites.

Wood's two basic constituents are the cellulose fibers and the natural plastic lignin. When the strong-in-tension but flexible cellulose fibers are cemented together in a matrix of lignin, a relatively strong and stiff composite material—wood—results. The same process occurs when a "mat" of fibers in a matrix of proteins and other organic substances forms the relatively strong and flexible composite material called leather.

Depending on the form of the constituents, natural and man-made composite materials can be classified as *fibrous, laminar,* or *particulate*.

Fibrous Composite Materials

Fibrous composites consist of at least two constituents—the fibers and the matrix. The fibers may be natural or synthetic and they may be metallic or non-

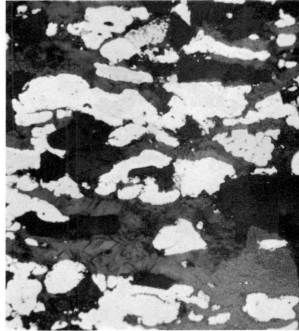

Bendix Corp.

7-11. Photomicrograph of a typical composite material called *Cerametalic.*® Used for friction clutches, this material consists of copper and alloys (shown as light areas in the photograph); graphite flakes (gray areas); and ceramic (dark areas).

metallic. Materials used for matrices are mostly synthetic resins, but metallic and some organic materials are also used. All materials are much stronger in fiber form than in bulk form. For example, commercially available glass fibers have 400,000 to 700,000 psi tensile strength, whereas common glass has only a few thousand psi tensile strength. The high strength-to-weight ratios achieved by fibrous composites make them ideal for such applications as aerospace and transportation where weight is a critical factor.

The properties of fibrous composites are influenced by:

◆ The types of fibers and their arrangement in the composite.

Technology of Industrial Materials

- The type of matrix.
- The manufacturing process employed to produce the composite.

Common fiber arrangements used in fibrous composites are *random mat, woven fabrics,* and *continuous* or *wound filaments.* The random mat arrangement is simple and is used to produce such fibrous composites as fiberglass-reinforced plastic films, synthetic leathers, and insulating materials. Fig. 7-12. One of the characteristics of these composites is that they have the same strength in all directions because the fibers have been randomly distributed in the mat. However, composites produced with fibers woven into a fabric have greater strength in the direction of the weave. The woven fabric method has been used for fiberglass-reinforced speedboat hulls, experimental houses, and tanks and other containers.

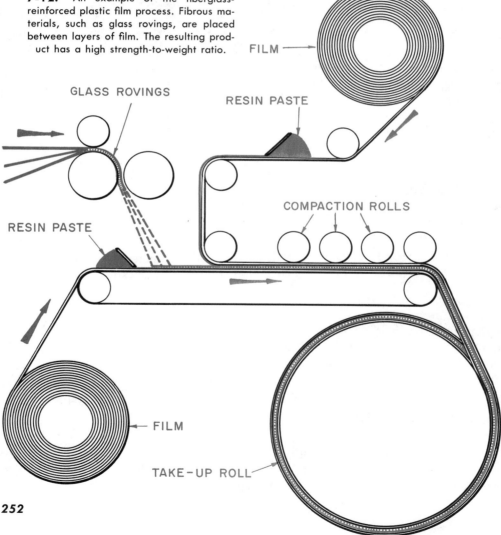

7-12. An example of the fiberglass-reinforced plastic film process. Fibrous materials, such as glass rovings, are placed between layers of film. The resulting product has a high strength-to-weight ratio.

Chapter 7. Miscellaneous Materials

If maximum strength in one direction is required, it can be achieved by making the composite with continuous filaments. The filaments are unidirectionally arranged parallel to each other like the wires in a cable to impart their maximum strength only in one direction. The continuous filament method has been used to produce metallic composites of extremely high strength and high temperature resistance. Continuous filaments can also be wound over a symmetrical mandrel at intersecting angles to produce composites of high strength. This method has been used for rocket boosters and tubular structural members requiring high strength and stiffness but extremely light weight. The matrix material can be applied to the fibers by spraying, brushing, dipping, or casting. Some of the advantages of fibrous composites are strength, stiffness, lightness, dimensional stability, and resistance to corrosion and heat.

Laminar Composite Materials

When two or more layers of similar or dissimilar materials are combined, the resulting material is a *laminar composite.* Laminar composites are among the oldest and best known of man-made composites. They have been used for over 3,000 years for such applications as swords and armor. Examples of common laminar composites are plywood, laminated lumber, formica tops, covers of hardbound books, and laminated plastics. The layers (lamina) involved in laminar composites may be of the same thickness throughout the entire composite, as in plywood, or of different thicknesses, as in formica tops and veneered furniture. In most laminar composites the layers are combined by adhesion; however, mechanical fastening is sometimes used. There is an almost infinite combination of materials that can be laminated such as metal to metal, metals to plastics, plastics to glass, leather to wood, and paper to gypsum.

There is a variety of reasons for producing laminar composites. Among the most important are that these materials can provide decoration and protection; strength; toughness and stiffness; wear, corrosion, and heat resistance; thermal, acoustical and electrical insulation or conductivity; color and light transmission; control of distortion; and production of unavailable structural shapes and sizes. Laminar composites are produced to increase the strength-to-weight ratios for airplanes and submerged vessels by combining strong and light materials. They are produced to increase resistance to wear, corrosion, and heat as in *clad* metals, in which the metal to be protected is overlaid with a thin layer of some highly resistant metal.

Laminar composites are also produced for safety purposes. For instance, in safety glass, plastic sheet and plate glass are combined to increase the impact strength and minimize shatter. Bimetallic laminar composites with different coefficients of expansion are used to regulate temperature fluctuation in refrigeration and heating units. (Fig. 1-3, page 15.) Composites in the form of "structural sandwiches" are used for increasing strength, stiffness, and heat or sound insulation as in carton boxes made of different papers and honeycomb structures made of aluminum for airplanes and space vehicles. Laminar composites are used for decoration in car interiors where synthetic and natural leathers or wood veneers are laminated to metal or plastic surfaces.

Particulate Composite Materials

When various-sized particles of a material are bonded together by a matrix, the

Technology of Industrial Materials

resulting material is called a *particulate composite*. A common example of a particulate composite is ordinary concrete (discussed in more detail in Chapter 5.) This material you will recall, consists of fine and/or coarse aggregate particles mixed in a portland cement matrix and hardened by chemical reaction with water. Particulate composites are produced and used more widely than fibrous and laminar composites. Materials difficult to combine by any other known process can be made into particulate composites having unique properties and applications. Successfully used combinations include metals in metals, metals in plastics, ceramics in metals, organic materials in inorganics and nonmetallic materials in other nonmetallics. The particles in these composites can be bonded together in a matrix as in concrete, or they can be sintered (heated without melting) as in construction bricks and some powder metallurgy products. Fig. 7-13.

In addition to portland cement concrete, several other particulate composites have been produced by bonding particles in a matrix of different material. Among the best known are asphalts, in which particles of sand and fine aggregate are bonded in a matrix of petroleum asphalt to form asphalt concrete used for road pavements. Metal powders have been bonded in a matrix of plastic material to produce cold solder and various types of bearing materials, among other items. Ceramic particles have been bonded in metal matrices to produce *cermets*. Cermets are particulate composites used for high-speed metal cutting tools, high-temperature valves, hot-drawing dies, and similar items. Combustible organic parti-

Amplex Division, Chrysler Corporation

7-13. Precision-finished machine parts produced by powder metallurgy.

Chapter 7. Miscellaneous Materials

cles have been bonded in inorganic matrices to form particulate composite materials used as solid fuels for rockets.

The best known materials produced by sintering particles together by heat into a cohesive, hard, and strong particulate composite are the *powder-metallurgy* products. Powder metallurgy is an important manufacturing process for producing metal-powder particulate composite articles. Fig. 7-14 illustrates the powder metallurgy process, which involves four basic steps:

1. Production of powders.
2. Blending and mixing of the constituents.
3. Compacting of the constituents into shapes.
4. Sintering the compact into a cohesive, hard, and strong particulate composite.

Metal powder can be produced by the following methods:

◆ *Reduction,* this is the most important method, used to convert metal oxides into powder by contact with a gas at temperatures below their melting point. Iron, molybdenum, cobalt, nickel, tungsten, and titanium powders can be produced by this method.

◆ *Electrolytic decomposition,* used for high-purity iron, copper, and silver powders.

◆ *Vapor condensation,* in which the metal is heated in a vacuum until it vaporizes, and the vapors are condensed into metal particles.

◆ *Atomization,* in which molten metal is dispersed into small particles by the impact force of high-velocity air or gas, as in metal spraying.

◆ *Milling* and *machining,* mechanical methods for producing metal powders by crushing and grinding metals.

◆ *Shotting,* which involves dropping molten metal through a fine screen into a coolant (water) to form round particles.

Blending and mixing of the constituents is accomplished by mechanical mixers, and its purpose is to secure a uniform

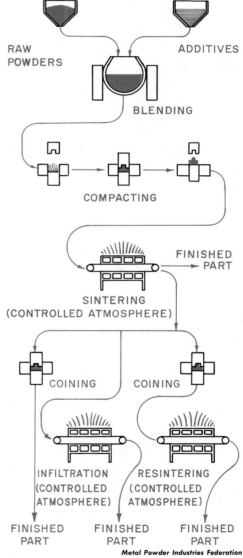

Metal Powder Industries Federation

7-14. Basic steps of the powder-metallurgy process.

255

7-15. Double-plunger hydraulic metal powder compacting press.

Baldwin-Lime-Hamilton Corporation

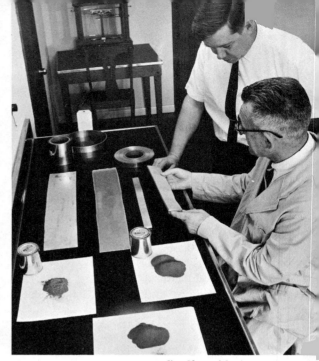

7-16. Various metal and alloy powders rolled into test strips.

Chas. Pfizer and Co., Inc.

distribution of the particles. The primary constituents in a metal-powder composite are the *metal powder or powders, lubricants* such as powdered graphite added to facilitate processing of the mix, and *other substances* as may be required to give certain properties to the composite. All the constituents are carefully weighed and placed in a rotating mixer until thoroughly mixed. They are then compacted. A variety of compacting methods may be used but the *close-die hydraulic pressing method* is the most common. Fig. 7-15. In this method a die of the same shape as the desired part is filled with the mix, and one or more hydraulically operated plungers compact the mix in the die by pressing it to about 20 to 60 tons per square inch. Some powder mixes, however, are extruded into desired shapes or rolled into sheet form. Fig. 7-16.

After the constituents have been compacted they are sintered by heating in a controlled-atmosphere furnace to about 70% of the melting temperature of the principal constituent. Fig. 7-17. The purpose of sintering is to enable the particles of the various constituents to bond together into a cohesive mass. The mechanism of sintering is not completely understood; however, scientists have offered the following possible explanations:

◆ Atomic diffusion of the various particles, resulting from physical contact and heat.

◆ Recrystallization as a result of high temperature.

◆ Growth and densification of grain.

Fig. 7-18 shows the structure of an iron-copper composite after sintering.

By proper mixing and sintering of metal powders with certain volatile materials, *skeletal particulate composites* can be produced. Skeletal composites consist of a sintered, strong, porous skeleton created by burning off the volatile materials dur-

Haller Division, Federal-Mogul Corp.

7-17. Continous-type sintering furnace, used for sintering transmission gears.

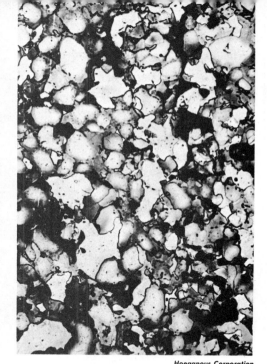

Hoeganous Corporation

7-18. Photomicrograph showing the structure of an iron-copper powder composite after sintering.

ing sintering. Such composites are used as porous material for metallic filters, Fig. 7-19, or they can be infiltrated with a metal of lower melting point by reheating the two metals one on the top of the other. As the metal with the lower melting point melts, it infiltrates the skeleton through capillary action and completely fills its pores. Or the skeleton may be impregnated by forcing another substance, such as oil, into the pores. This process is used to produce self-lubricating bearings. Fig. 7-20.

Amplex Division, Chrysler Corp.

7-19. Porous particulate composites used for gas and liquid filters

Amplex Division, Chrysler Corp.

7-20. Typical self-lubricating bearings.

257

LUBRICANTS

When two contacting surfaces are sliding (rubbing) in different directions, a force called *friction* results, opposing the motion of the sliding surfaces. Friction is caused by many factors including the *irregularities* (roughness) of the sliding surfaces, and the load carried by the two surfaces.

A substance can be placed between the sliding surfaces to reduce friction and minimize wear. Fig. 7-21. Such a substance is called a *lubricant*. More specifically, a lubricant is any liquid or solid placed between two sliding surfaces for one or more of the following purposes:

- Reduce friction and wear by minimizing direct contact between the sliding surfaces.
- Cool the sliding surfaces by removing the generated heat of friction.
- Clean the sliding surfaces by carrying away particles of contaminating substances.
- Seal the gap between the sliding surfaces to prevent loss of pressure.
- Protect the sliding surfaces from oxidation.

Many types of lubricants have been developed. On the basis of origin, composition, consistency, application, and method of manufacturing, lubricants can be classified in five groups:

- Petroleum lubricants.
- Synthetic lubricants.
- Fatty (fixed) oils.
- Lubricating greases.
- Solid (or extreme pressure) lubricants.

Petroleum Lubricants

Most of the petroleum lubricants used today are produced by refining crude petroleum. Crude petroleum is the result of longtime, low-temperature gradual decomposition of marine organic matter (fossils). Crude petroleum is a mixture of hydrocarbons—molecules containing carbon and hydrogen. Sulfur, nitrogen, and some other chemical compounds are also present as impurities in crude petroleum. The three predominant types of hydrocarbons which form crude petroleum are *paraffins*, *naphthenes*, and *aromatics* (so named because formerly they were distinguished from other compounds by their characteristic odors rather than by their structures). Thus crude petroleum and the lubricants derived from it are classified in these four groups:

- Paraffinic.
- Naphthenic.
- Aromatic.
- Mixed.

Paraffin-based petroleums consist primarily of paraffins with an appreciable amount of petroleum wax. They have relatively low specific gravity, viscosity, and sulfur content, and are found in Pennsylvania, West Virginia, and other Eastern states. Naphthene-base petroleums consist primarily of naphthene with no petroleum

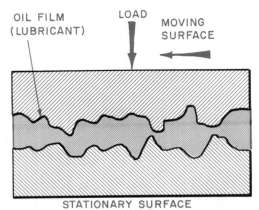

7-21. Schematic showing friction due to surface irregularities and load, and the separating effect of the lubricant between two rubbing surfaces.

wax. They have relatively high specific gravity, varying amounts of sulfur content, and are found in Kansas, Oklahoma, and Texas. Aromatic petroleums are found in California. Mixed-base petroleums consist primarily of paraffins and naphthenes with relatively high sulfur content and small amounts of petroleum wax.

The crude petroleum extracted from underground by drilling is transported to the refinery by pipe systems or by tankers. The refining of crude petroleum is a very complex thermochemical process called *distillation*. At the refinery, Fig. 7-22, (or sometimes earlier) certain chemical compounds are added to the crude to modify certain properties and enhance the distillation process. These compounds, called *additives*, include neutralizers, corrosion inhibitors, and anti-fouling agents.

The petroleum is heated in vacuum until it vaporizes, and the vapors are condensed into such classes of petroleum products as gases, gasolines, gas oils, lubricant distillates, and residues. Lubricating oils are produced by further treating the lubricant distillates and residues as

The Standard Oil Company

7-22a. Refining of petroleum. Partial view of a modern refinery.

7-22b. Central control room of the refinery.

The Standard Oil Company

259

Technology of Industrial Materials

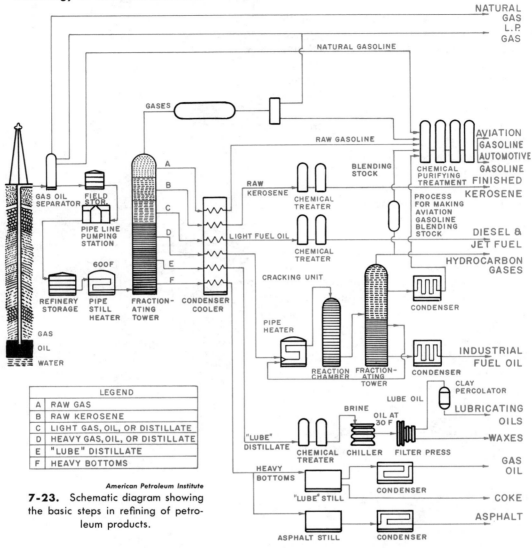

American Petroleum Institute

7-23. Schematic diagram showing the basic steps in refining of petroleum products.

shown in Fig. 7-23. The oils are treated by solvent refining and solvent dewaxing to remove such unstable and unwanted constituents as petroleum wax. Fig. 7-24. Because carbon and hydrogen may exist in many combinations in the hydrocarbon molecules of lubricating oils, different grades of oils having various properties can be produced. Such grades as light spindle oils, intermediate lubricating oils, and heavy oils, with varying viscosities and boiling ranges, are produced.

Synthetic Lubricants

A synthetic lubricant is an oil produced by chemical processing of various organic and inorganic materials. Synthetic lubricants have been developed for applications in which petroleum oils could not function satisfactorily. However, due to

7-25. Modern continuous lubricating-grease processing unit used to produce lithium-, calcium-, and sodium-based greases.

gears. The lubricating properties of greases are due primarily to the lubricating oil component, with the soap component and additives imparting other desirable properties. Depending on the type of soap base used, greases vary in consistency from free-flowing liquid to solid. Greases can be classified in a number of ways, but in terms of the soap base they are classified as calcium, sodium, lithium, aluminum, barium, multi-soap, nonsoap, and asphalt. Fig. 7-25. Table 7-G summarizes the composition, properties, and typical applications of greases.

Another classification system of greases has been developed by the National Lubricating Grease Institute (NLGI) which designates greases by a number based on consistency. The consistency test, standardized by ASTM, is described in Chapter 8.

Solid Lubricants

Solid lubricants have been developed for use in applications where neither lubricating oils nor greases could be used successfully. Some solid lubricants used today are graphite, molybdenum disulfide, lead, and silicons. These lubricants can be used in extreme pressure (high load-carrying bearings) and temperature applications such as in mills for rolling steel and in hypoid gears. They can be used in powder form or as suspensions in a semi-liquid form.

FUELS

Any material which exists in sufficient quantities to be extracted or manufactured economically and which possesses sufficient energy (chemical or nuclear) to be released and used in the form of *heat*

Technology of Industrial Materials

Table 7-G

Composition, properties, and uses of principal types of greases.

Composition	Properties	Typical Uses
Calcium-Soap-Based Greases		
Calcium soap, lubricating oil, and some additives. These are the most common and least expensive greases, known as low-temperature greases.	Good water resistance; approximate ASTM dropping point 160° F; fair resistance to breakdown by working; smooth to buttery texture.	General purpose industrial uses; for cup and pressure gun use; for centralized lubricating systems but not recommended for working temperatures above 175° F.
Sodium-Soap-Based Greases		
Soda soap, lubricating oil, and some additives; known as high-temperature greases.	Approx. ASTM dropping point 325° F; excellent resistance to break-down by working; good working performance at 300–350° F temperatures; fibrous to buttery texture; water soluble, but of high melting point.	For anti-friction, plain bearings, and roll and ball bearings; for moderately high temperature lubrication, etc.
Lithium-Soap-Based Greases		
Lithium soap, lubricating oil, and additives; known as multi-purpose greases.	Good mechanical stability; excellent resistance to breakdown by working; high temperature and water resistance; approx. ASTM dropping point 350° F; smooth to buttery texture.	For uses with wide temperature variations (-100 to $+300°$ F); for aircraft and similar applications.
Aluminum-Soap-Based Greases		
Aluminum soap, lubricating oil, and additives; but expensive.	Approx. ASTM dropping point 185° F; good water resistance, but poor resistance to break-down by working; high adhesive characteristics; smooth to stringy but not fibrous; brilliant and transparent in appearance; sticky.	For cams, chains, oscillating bearings, and where extreme tackiness is needed.
Barium-Soap-Based Greases		
Barium soap, lubricating oil, and additives; relatively high soap content; multi-purpose greases.	Approx. ASTM dropping point 350° F; good water resistance, but poor resistance to break-down by working; relatively high melting point; smooth or fibrous.	For all bearings; chassis; general automotive lubrication, but not for sub-zero temperature applications.
Multi-Soap-Based Greases		
Calcium-sodium soaps, lubricating oil, and additives mixed to produce greases of general purpose characteristics.	Good speed characteristics; other properties depend on type of mix.	For life-lubricated bearings and multi-purpose uses.

Chapter 7. Miscellaneous Materials

Table 7-G (Continued)
Composition, properties, and uses of principal types of greases.

Composition	Properties	Typical Uses
Nonsoap-Based Greases		
Silica gel and bentonite clay; known as dispersed or microgel greases.	Have no specific melting point; excellent wear-protection characteristics; high temperature resistance; good mechanical stability but poor corrosion protection.	For steel mills and textile mills; uses requiring freedom from drip at high temperatures; for plain bearings subjected to high temperature.
Asphalt (Petroleum)-Based Greases		
Blends of asphaltic residue from petroleum distillation, mixed with oil, graphite, mica, and fat.	Good adhesive characteristics; poor temperature characteristics; high internal friction and low melting point.	For large, slow-speed open gears and for dressing of wire rope.

is a *fuel*. Fuels produce heating energy by combustion in order to provide the power needed for running internal combustion engines, jet engines, nuclear reactors, generators of electricity, and heating and refrigeration machinery, to mention only a few uses.

The amount of heating energy released by burning a unit amount of a fuel (pound or cubic foot) is its *heating value* (HV), and it is the most important property of the fuel. The HV of fuels is expressed in *British thermal units* (Btu) or in the metric system, in *calories* (cal). One Btu is the amount of heat required to raise the temperature of one pound of water one degree Fahrenheit, while one calorie is the amount of heat required to raise the temperature of one gram of water one degree centigrade.

The HV of fuels is determined by using an apparatus called a calorimeter. One of the standard types, the oxygen-bomb calorimeter, is shown in the next chapter, Fig. 8-42.

Some fuels are manufactured while others are natural, but all fuels are classified as gaseous, solid, or liquid.

Gaseous Fuels

These are fuels which, at normal temperatures and pressures, are in a gaseous state. Most such fuels are mixtures of different gases; their properties depend on the number and percentage of the constituents present in the mix.

Gaseous fuels are considered "ideal fuels" because they can be stored, distributed, ignited, and their combustion can be controlled, all with relative ease. They burn cleanly and without requiring complicated and expensive equipment. However, some of these fuels are toxic and relatively expensive.

Gaseous fuels are classified as either *natural* or *manufactured.* Among the natural gaseous fuels, the most important is the one known simply as natural gas. In the manufactured category are coal and coke gases, blast furnace gas, producer gas, water gas, oil gas, sewage gas, and—the most important one—liquified petroleum. In fact, because the use of liquid petroleum gas and natural gas has become so widespread, most of the other manufactured gases are diminishing in importance as fuels. Especially in recent years, the

Technology of Industrial Materials

Table 7-H

Types, composition, properties, and uses of gaseous fuels.

Origin and Composition	Properties	Typical Applications
Natural Gases		
Naturally occurring gas from petroleum wells. Methane, ethane, and propane form the three principal combustibles, with minor quantities of carbon dioxide, helium, and nitrogen.	In gaseous state in a relatively wide temperature range; no characteristic odor; heating value range from 850 to 1950 Btu/cu. ft.	Relatively inexpensive; used for many domestic commercial and industrial applications.
Coal and Coke-Oven Gases		
Produced by heating coal or as a by-product of coke-ovens at steel mills. A mixture of methane and hydrogen with some sulfur compounds present.	A relatively low heating value, from 500–600 Btu/cu.ft.	Used at the source of production and for many domestic applications for heating, refrigeration, etc.
Blast Furnace Gas		
Produced by partial combustion of coke in a blast furnace. A mixture of methane, carbon monoxide, nitrogen, and hydrogen.	A relatively low heating value, from 80–150 Btu/cu. ft.	Used to operate gas engines to produce electricity for the steel mill and for heating and steam generating.
Producer Gases		
Produced in gas generators by blowing air and steam through a bed of coke, coal, or wood fire, resulting in carbon monoxide. Mixture of carbon monoxide and nitrogen.	A relatively low heating value, from 120–180 Btu/cu. ft.	Expensive and used for such applications as in steel mills, glass and ceramic plants, etc.
Water Gas		
Produced in gas generators by blowing air first and steam second through a bed of coal, coke, or wood fire. Mixture of hydrogen and carbon monoxide.	Heating value range is from 525–555 Btu/cu. ft.	For domestic uses, for industrial uses, and as a mixing agent in other gases.
Liquefied Petroleum Gases (L.P.G.)		
Produced from petroleum hydrocarbons that can be liquefied at relatively low pressures. Mixture of methane, ethane, propane, and butane.	Relatively high heating value range, from 1000–4000 Btu/cu. ft.	Expensive; used in domestic applications; industrial uses for heating and power; for industrial chemical uses, especially production of synthetic rubber.

Chapter 7. Miscellaneous Materials

Table 7-K
Selected examples of additives used in diesel fuels.

Additive Type	Chemical Composition	Purpose
Cetane number improver	Alkyl nitrates	Upgrade ignition quality.
Stabilizer-dispersant	Alkaline metal sulfonates	Decrease rate of particle formation.
Pour depressant	Olefinic-type polymers	Improve pour point.
Corrosion inhibitor	Ammonium sulfonates	Provide protection against rust.
Metal deactivator	Disalicylidene propylene diamine	Arrest filter screen plugging.
Odor improvers	Aldehydes	Improve odor of exhaust.

Texaco Inc.

of low economic value (as are found in the distillates and residues) into hydrocarbons of smaller molecules and of high economic value (such as those which make up gasolines). Cracking is accomplished either by heat and pressure (*thermal cracking*) or by chemical catalysts (*catalytic cracking*) and is performed in complex equipment called *crackers*. Fig. 7-26 shows a typical cracker for fluid catalytic cracking of gasoline. Not only gasoline but also coke, tar, and gum are produced during cracking of petroleum distillates and residues. Thermal cracking of distillates and residues gives lower yield of gasoline production and produces gasolines of lower anti-knock and higher gum-forming characteristics than catalytic cracking. While cracking breaks down hydrocarbons of larger molecules, polymerization and alkylation combine smaller molecules of gaseous hydrocarbons into larger molecules and increase the yield of gasoline production from crude petroleum.

The most common method used to distinguish one liquid fuel from another is based on the boiling range of the fuel. For example, diesel fuels have a relatively higher boiling range than gasolines.

Diesel Fuels

Diesel fuels are liquid fuels used for diesel internal-combustion engines to provide power in such applications as locomotives, trucks and buses, boats, ships and submarines, tractors, heavy-construction equipment, and stationary engines, among others. Diesel fuels are produced from petroleum distillates or residues and are relatively less expensive than gasolines. Due to the wide variety of diesel engine designs, diesel fuels vary widely in composition and properties, ranging from heavy oils such as kerosene to very light fuels such as low-grade gasolines. These fuels are burned in the compression chamber of the diesel engine, and the manner in which they ignite and burn affects the operation and efficiency of the engine. Therefore the *ignition quality* of the fuel is among its most important properties.

The ignition quality of a diesel fuel depends on its composition, especially the structure of the hydrocarbon molecule, and is specified in terms of a *cetane number*. The cetane number is an index indicating the time required for ignition of the fuel in the compression chamber of the engine when it reaches a critical temperature (the *spontaneous ignition temperature*). The test for determining the cetane number has been standardized and is discussed in Chapter 8.

The characteristics of diesel fuels can be improved by using additives. Table 7-K

Table 7-L
ASTM standard specifications for fuel oils.

Grade of Fuel Oil[a]		Flash Point, Deg. F (Deg. C)	Pour Point, Deg. F (Deg. C)	Water and Sediment, Percent by Volume
		Min	Max	Max
No. 1	A distillate oil intended for vaporizing pot-type burners and other burners requiring this grade of fuel	100 or legal (38)	0	trace
No. 2	A distillate oil for general purpose domestic heating for use in burners not requiring No. 1 fuel oil	100 or legal (38)	20[b] (−7)	0.10
No. 4	Preheating not usually required for handling or burning	130 or legal (55)	20 (−7)	0.50
No. 5 (Light)	Preheating may be required depending on climate and equipment	130 or legal (55)	...	1.00
No. 5 (Heavy)	Preheating may be required for burning and, in cold climates, may be required for handling	130 or legal (55)	...	1.00
No. 6	Preheating required for burning and handling	150 (65)	...	2.00[e]

[a] Recognizing the necessity for low-sulfur fuel oils used in connection with heat-treatment, nonferrous metal, glass, and ceramic furnaces and other special uses, sulfur requirement may be specified as follows: for grade No. 1, 0.5% sulfur maximum; for No. 2, 0.7% sulfur maximum; and for grades 4, 5, and 6, no limit on sulfur. Other sulfur limits may be specified only by mutual agreement between the purchaser and the seller. It is the intent of these classifications that failure to meet any requirement of a given grade does not automatically place an oil in the next lower grade unless in fact it meets all requirements of the lower grade.
[b] Lower or higher pour points may be specified whenever required by conditions of storage or use.

gives selected examples of additives used in diesel fuels.

Fuel Oils

Fuel oils are produced as "straight run" oils during the distillation of crude petroleum and by cracking of petroleum distillates and residues. They vary widely in composition and properties, ranging from very light oils requiring no preheating for burning to very heavy oils which do require preheating for burning and handling.

Fuel oils are used extensively for domestic and commercial heating applications and in industry for heating and for generation of electricity and steam as well as in many metallurgical melting and re-

Carbon Residue on 10 Percent Bottoms, Percent	Ash, Percent by Weight	Distillation Temperatures, Deg. F (Deg. C)		Saybolt Viscosity, sec.		Gravity Deg. API	Copper Strip Corrosion
		10 Percent Point	90 Percent Point	Universal at 100° F (38° C)			
Max	Max	Max	Min	Max	Min	Max	Min	Max
0.15	...	420 (215)	...	550 (288)	...	...	35	No. 3
0.35	...	f	540c (282)	640 (338)	(32.6)d	(37.93)	30	...
...	0.10	...	...	...	45	125	...	...
...	0.10	...	...	...	150	300	...	...
...	0.10	...	...	...	350	750	...	...
...	...	...	...	...	(900)	(9000)	...	...

cWhen pour point less than 0° F is specified, the minimum viscosity shall be 1.8 cs (32.0 sec., Saybolt Universal) and the minimum 90% point shall be waived.
dViscosity values in parentheses are for information only and not necessarily limiting.
eThe amount of water by distillation plus the sediment by extraction shall not exceed 2.00%. The amount of sediment by extraction shall not exceed 0.50%. A deduction in quantity shall be made for all water and sediment in excess of 1.0%.
fThe 10% distillation temperature point may be specified at 440° F (226° C) maximum for use in other than atomizing burners.

fining applications. To burn efficiently, most of these fuels require atomization by special burners whose design depends on the viscosity of the fuel. The heating values of the fuel oils vary, with those produced by cracking having relatively higher densities and heating values than those produced as "straight runs" from distillation of crude petroleum. The ASTM has standardized the specifications for fuel oils; these are given in Table 7-L.

Gas Turbine or Jet Fuels

Gas turbine fuels are often called jet fuels or aviation turbine fuels. Although some gas turbines still use solid fuels in powder form, most now use petroleum and synthetic-based liquid fuels. Aviation turbines (jet engines) use a variety of liquid fuels, ranging from low-grade gasolines to kerosenes, depending on the design and application of the engine. Due to the design of turbines, which is quite different from the design of gasoline and diesel engines, the requirements for jet fuels are relatively less exacting than those for gasoline and diesel fuels. Most turbine fuels are blends of petroleum and synthetic hydrocarbons which possess good *volatility*. Volatility means the tendency to vaporize readily. Good fuel volatility is needed to start the engine when it is cold, but to prevent vapor lock the fuel should not vaporize at high temperatures. Vapor lock occurs when the fuel vaporizes in the lines, reducing the amount of fuel delivered in the combustion chamber of the engine.

The ASTM standardized specifications for aviation turbine fuels are given in ASTM *Standard* D1655. According to ASTM, three types of aviation turbine fuels are available:

Jet A—A relatively high-flash-point distillate of the kerosene type. (Flash point is the temperature at which the fuel vaporizes enough to ignite in the presence of a flame.)

Jet B—A volatile distillate with relatively wide boiling range.

Jet A-1—A kerosene type similar to type Jet A but incorporating special low-temperature characteristics for certain applications.

Gasolines

Based on the volume consumed annually, gasoline is the most important liquid fuel. Basically it is a complex mixture of volatile hydrocarbons having a wide boiling range and vapor pressures. It can be produced as *straight run* gasoline from the primary distillation of crude petroleum or as *cracked* gasoline by cracking the petroleum distillate. Most gasoline grades used today are blends of straight run and cracked gasolines and various additives. Gasoline has a relatively lower boiling range and is more volatile than diesel fuels, turbine fuels, and fuel oils. Among the important properties of gasoline which affect engine performance are *volatility, heating value,* and *ignition quality*—the manner in which it burns in the combustion chamber of the engine.

Volatility is an extremely important property of gasolines. It is determined by the *ASTM Standard distillation test* and/or the vapor pressure test as explained in the next chapter. The heating value of gasoline is determined by the calorimeter method discussed under *Testing of Solid Fuels,* also in the next chapter. The manner in which the gasoline burns in the combustion chamber of the engine is related to, among other things, the knock characteristics of the fuel. Knocking refers to the characteristic rattling noise developed in the combustion chamber as a result of "abnormal" burning of the fuel. The *octane number* of a gasoline is its knocking index. This number is determined by using a standard test engine which has variable compression and which operates on a mixture of isooctane and heptane. Isooctane knocks less than any gasoline, while heptane knocks more than any gasoline. Therefore when a mixture of isooctane and heptane *matches* the knock of a tested gasoline specimen, the percentage of isooctane in the mixture indicates the octane number of that gasoline. An octane scale from 0–100 is used for automotive gasolines, with some grades of aviation gasoline produced with octane numbers higher than 100.

Chapter 8 Testing of Industrial Materials

NATURE AND SCOPE OF MATERIALS TESTING

Among the most important industrial uses of materials are their applications in building equipment and structures to perform specified functions under given conditions. For example, an airplane is built for transporting people and products by air; a bridge for crossing a river; a submarine for underwater travel; and a space vehicle for outerspace travel. As explained in Chapter 1, to obtain the best performance from equipment and structures, the behavior of various materials under service conditions must be known. The main purpose of materials testing is to study the behavior of materials under specified conditions. This chapter will deal with many tests and testing procedures developed for that purpose.

Materials testing can be classified as follows:

◆ *Destructive* or *nondestructive*. In destructive testing the specimen is damaged so that it cannot be used again. Destructive testing is mostly done to determine the mechanical properties of materials under loads. Nondestructive testing does not damage the specimen and is commonly done to locate defects in materials and products. Nondestructive testing is frequently associated with *inspection* of materials and products.

◆ *Static* and *dynamic*. In dynamic testing, loads are applied and then immediately removed, as by a hammer blow. Appreciable impact force is usually involved.

Most material testing, however, is performed under static loads. Such loads are allowed to remain on the specimen, usually for just a short time, and are often applied in such a way that impact force is negligible. This type of testing is referred to as static testing.

◆ *Room-temperature, high-temperature,* or *low-temperature*. Although some tests are performed at either sub-zero or high temperatures, most are done at room temperature.

◆ *Field* or *laboratory*. Field tests are conducted under actual service conditions, often on a worksite. Laboratory tests involve simulating one or more service conditions, usually in a setting where results can be mostly closely observed and measured.

Before describing the various tests and testing procedures, some attention should be given to such factors as:

◆ Characteristics of specimens to be tested.
◆ Characteristics of tools and equipment required.
◆ Economy of testing.
◆ Accuracy required.
◆ Testing procedures.

The size,* shape, and composition of a specimen are its basic characteristics; as

---

*Specimen sizes and other testing data in this chapter are often given in Customary units. Listing the metric equivalents would be of little value because in the metric system the standards themselves are different. In other words, a 2″ specimen will probably not be replaced by a 50.8mm specimen but by one of a more convenient length, perhaps 50mm. Standards to be used in the United States have not been determined as this book is prepared for the press.

Technology of Industrial Materials

these vary, procedures for testing the specimen will also vary. Some specimens are unfinished pieces of material, such as wood as it comes from the lumber mill or a piece of steel bar as it comes from the steel mill. Other specimens may have to be finished, as by machining or grinding. In some testing situations one specimen may be sufficient to provide reliable results; in other cases several specimens may be required.

For many tests the necessary equipment, including tools and fixtures, is available from commercial sources. Sometimes, however, special tools and fixtures must be built. In many testing situations, the expense involved in performing a test and the accuracy required should be given special attention before deciding on a particular test.

Materials testing requires that standard procedures be followed as much as possible to assure reliability and validity. Standardized testing procedures are recommended and published by the American Society for Testing and Materials (ASTM) in the *ASTM Standards*. This technical, scientific, and educational society is responsible for the standardization of materials specifications, methods of testing, and definitions in the field of engineering materials. Most of the specifications, methods of testing, and definitions included in the *ASTM Standards* are extensively used by industry and by professional technical organizations throughout the United States. Therefore a copy should be available in the materials testing laboratory for student as well as teacher reference.

FAILURE IN MATERIALS

To determine the behavior of materials under specified conditions, an understanding of the general ways in which materials fail is necessary. A material (or a product) has failed if it no longer performs the service functions for which it was intended. Any material can fail under adverse chemical or physical conditions, or a combination of the two.

- Materials may fail because of *chemical reactions* resulting in rusting or corroding.
- *Physically*, materials may fail because of excessive loads, resulting in cracking, breaking, splitting, crushing, buckling, tearing, abrading, or wearing. Such failure is particularly important among industrial materials.

Deformation of a material is dimensional change. Before failing under an

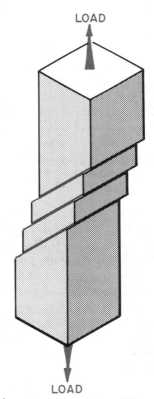

8-1. Schematic representation of the process of slip in metallic material subjected to tensile stresses.

excessive load, a material tends to resist deformation. The resistance of a material to failure by deformation is called its *strength*; this resistance is primarily caused by the interatomic forces of the material as explained in Chapter 2.

In metallic materials the principal process of deformation is *slip*. Slip is the "sliding" of atomic planes in the space lattice (see Chapter 2) of the material due to excessive applied force. Fig. 8-1. The fracture induced by slip may occur suddenly, not preceded by any appreciable deformation, as in the case of brittle metallic materials (brittle fractures); or it can occur slowly, followed by a noticeable deformation as in the case of ductile materials (ductile fractures). If a fracture in a metallic material cuts across metallic grains, it is referred to as *transcrystalline,* but if it occurs at the boundaries of metallic grains, it is referred to as *intergranular.* Fig. 8-2.

Chapter 8. Testing of Industrial Materials

TYPES OF TESTS— DESTRUCTIVE

Among the tests selected for discussion in this chapter are:

- Tensile and compression.
- Hardness.
- Bending (flexure).
- Impact.
- Fatigue.
- Creep, torsion, and shear tests.

Most of these tests have been standardized by ASTM and are found in the *ASTM Standards.* A distinction should be made here between *testing* and *inspection.* *Testing* usually refers to operations performed for the purpose of collecting data to determine, among other things, the properties of materials. *Inspection,* on the other hand, usually refers to the examination of materials (or products) to locate flaws, defects, or undesirable qualities. As mentioned earlier, inspection is usually nondestructive and is based on the analysis of visual, electrical, radiographic, magnetic, ultrasonic, and laser-beam phenomena. Some of these inspection tests have been standardized by ASTM.

Tensile and *compression* tests are two of the most important and most commonly performed tests. Their purpose is to provide data to study the behavior of materials under tensile and compressive loads. Though different, these tests are often discussed together because of certain common elements. Tensile and compression tests require the specimen to be subjected to uniaxial tension or compression loading (loading along the central axis of the specimen) until it fractures. Fig. 8-3 illustrates the basic principle of uniaxial

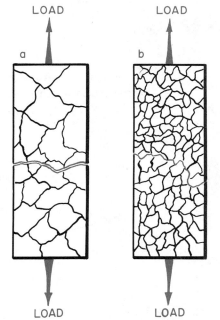

8-2. Schematic representation of fractures in metallic materials: (a) transcrystalline and (b) granular.

Technology of Industrial Materials

tensile and compression loading of a specimen. In the case of tensile tests, the specimen is gripped from its two ends and pulled apart. In compression tests the specimen is compressed perpendicularly until it fractures. Tensile and compression tests are used with such materials as ceramics, plastics, wood, metals, fibers, paper, and leather, among others.

Tensile Tests

Most materials used in manufacturing and construction can be tested in tension. The ASTM has standardized tensile tests for steel products, cement and concrete, rubber, wood, plastics, and insulating materials. Recommended standard designs of specimens as well as procedures for tensile tests of various materials are given in *ASTM Standards*. Tensile tests provide data on basic properties of materials—data that can be applied directly to design problems. However, the main purpose of tensile tests is to determine the behavior of materials under *loading* (stress-strain) conditions. *Stress* is the force per unit of area; it is measured in pounds per square inch (psi). In the metric system, stress is measured in kilograms per square millimeter (kg/mm^2)*. *Strain* is the change in the original size of a specimen due to force and is measured in inches of change per inch of the original dimension (in/in).

A useful method of representing the behavior of a specimen under tensile load is to use a stress-strain curve or diagram. This type of diagram can be plotted for any such specimen. Because of its unique structure each material has its own curve. Fig. 8-4.

The following procedure may be used to plot a stress-strain curve:

1. On the vertical axis (ordinate) of a sheet of graph paper locate stress points at equal intervals for each 1000 psi.
2. Measure the strain at each stress point and enter it on the horizontal axis (abscissa).
3. Draw perpendiculars from each stress point and corresponding strain point.
4. Connect the various intersections with a curve as in Fig. 8-4.

Among the properties directly observed (and those that can be estimated) in tensile tests are:

- Proportional limit.
- Yield point and yield strength.
- Ultimate tensile strength.
- Fracture (or breaking) strength.
- Ductility, elongation, and reduction of area.
- Modulus of elasticity.

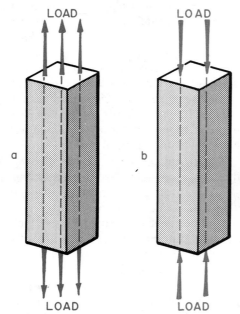

8-3. Specimens subjected to uniaxial stresses: (a) uniaxial tensile stress and (b) uniaxial compressive stress.

*For more information on the metric system of measurement see Appendix D.

Chapter 8. Testing of Industrial Materials

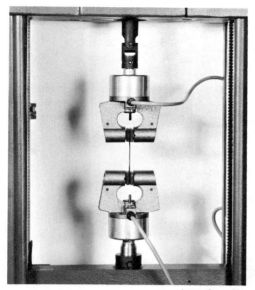

8-9c. Pneumatic grips for tensile testing of plastics.

8-9d. Grips for tensile testing of rubber.

8-9e. Grips for tensile testing of wire and rope.

flexible metallic tube) is probably the most common. Fig. 8-10. In this method as load is applied to the specimen, pressure is placed on a hydraulic capsule. This pressure is transmitted to the Bourdon tube, causing distortion of the tube. This distortion is registered by the measuring unit, thus indicating the load.

A testing machine must be carefully calibrated to assure accuracy of the test. Although all machines are factory calibrated, a check after installation in the laboratory and periodic checks thereafter are recommended. Several methods are recommended by the ASTM to calibrate testing machines. A convenient but not highly accurate method is to calibrate the machine by using a specimen of known tensile strength.

Most testing machines are equipped with certain attachments designed to increase the versatility and accuracy of the machine. Some of these attachments are strain-measuring extensometers for measuring the amount of strain on the specimen; stress-strain recording systems for plotting the stress-strain curve; automatic program attachments for automatic testing; center punches for setting off the gage

Technology of Industrial Materials

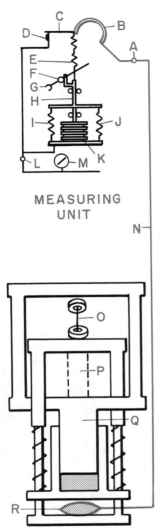

8-10. Schematic representation of a load-measuring system used with hydraulic tensile testing machines. (A) Connection from capsule. (B) Bourdon tube. (C) Baffle. (D) Air jet. (E) Iso-elastic springs. (F) Rack and pinion. (G) Pointer on dial face. (H) Slide bearing assembly. (I) Pneumatic servo motor. (J) Power springs. (K) Syphon bellows. (L) Orifice. (M) Gauge. (N) Line from measuring unit to tensile testing machine. (O) Tension specimen. (P) Compression specimen. (Q) Cylinder with piston and hydraulic liquid. (R) Capsule with hydraulic liquid.

length marks on the specimen; percent gage for direct indication of percentage of elongation; reduction of area gage; and attachments for compression, bending, shear, and hardness tests. Some of these attachments come as standard equipment with the machine; others must be built or purchased independently.

Testing Procedure

Before a test begins, the steps of procedure should be outlined. The ASTM recommends specific procedures for most standardized tests; these procedures can be adapted to suit specific situations.

To record data as accurately as possible, it is desirable that appropriate tables, graphs, and charts be made before the test begins.* For samples of data-recording tables and charts see Appendix I. To minimize the possibility of introducing non-axial forces on the specimen, which may affect the results, an attempt should be made to place the specimen on the testing machine in a way that will pull or compress it uniaxially. When wedge-type grips are used, all of the grip, or at a minimum 75% of it, should be in contact with the specimen.

Speed of testing or rate of applying the load on the specimen should be carefully determined. The speed of testing varies for different materials and also can be different before and after the yield point. For metallic materials, speeds of 0.125" per minute up to the yield point and 0.5" per minute above the yield point are recommended. Rates of applying loads for other materials are specified in the *ASTM Standards*.

---

*Note to the instructor: It is advisable that two students be assigned to perform a tensile test—one to operate the controls and the other to record the data. This procedure not only facilitates accurate data recording but also reduces the risk of accidents.

Chapter 8. Testing of Industrial Materials

Compression Tests

Compression testing is the opposite of tensile testing. However, many of the basic principles discussed in connection with tensile testing apply to compression testing. The universal testing machine used for tensile tests is also used for compression tests. On most of these machines, for compression testing the specimen is placed between the lower head and the table. The load is applied through the base. The table moves upward toward the lower head which is fastened at a stationary position. Fig. 8-8.

Although compressive strength is the most commonly determined property in commercial testing, other properties such as yield strength, yield point, and modulus of elasticity are also determined. Compressive strength can be determined by dividing the maximum load by the original cross-sectional area of the specimen. However, due to the different behavior of ductile and brittle materials under compressive loads, it is necessary to study these materials separately. The compressive strength of brittle materials, which fracture under compressive loads, can be given a numerical value. This is determined by dividing the breaking load by the original area of the specimen. Examples of brittle materials are cast iron, wood, concrete, brick, and glass. The formula for finding compressive strength is:

$$\frac{\text{Compressive}}{\text{Strength (psi)}} = \frac{\text{Breaking Load (lbs.)}}{\text{Original Area (sq. in.)}}$$

It is difficult to determine the compressive strength of ductile materials because they do not fracture under compressive loads. Numerical values for compressive strength of ductile materials are indefinite—that is, they cannot be determined because the diameter of the specimen under loads tends to increase gradually.

Table 8-A
Recommended solid round metallic specimens.

Specimens	Diameter in Inches	Length in Inches
Short	$1\frac{1}{8} \pm 0.01$	1 ± 0.05
Medium length . .	$\frac{1}{2} \pm 0.01$	$1\frac{1}{2} \pm 0.05$
	0.80 ± 0.01	$2\frac{3}{8} \pm \frac{1}{8}$
	1 ± 0.01	$3 \pm \frac{1}{8}$
	$1\frac{1}{8} \pm 0.01$	$3\frac{3}{8} \pm \frac{1}{8}$
Long	0.08 ± 0.01	$6\frac{3}{8} \pm \frac{1}{8}$
	$1\frac{1}{4} \pm 0.01$	$12\frac{1}{2}$ min

In order to reduce inaccuracies in results, care should be exercised in selecting and preparing the specimen and performing the test in a correct manner. The ASTM recommends that whenever possible, compression specimens should be made in solid cylindrical shapes. This general rule is more applicable to some materials, such as cast iron and concrete, than to others, such as rubber and paper. In preparing compression specimens, the ratio of length to diameter (L/D) is a critical factor and should be considered carefully. Based on L/D ratio, the ASTM recommends three types of specimens for metallic materials:

◆ *Short-length specimens* are those with length equal to nine-tenths the diameter (L = 0.9D). This type of specimen is recommended for bearing metals and similar applications.

◆ *Medium-length specimens* are those with length equal to three diameters (L = 3D). These specimens are usually employed to determine the compressive strength of metallic materials.

◆ *Long specimens* are those with length equal to eight to ten diameters (L = 8D to 10D). Long specimens are used to determine the modulus of elasticity in compression of metallic materials. Table 8-A gives the dimensions of metallic specimens as recommended by ASTM.

Technology of Industrial Materials

8-11a. Compression testing of concrete specimen. The specimen has just fractured.

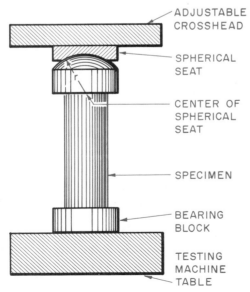

8-11b. Spherical attachment used for compressive tests as recommended by ASTM.

Though the dimensions are recommended for metallic specimens, the actual sizes of specimens for other materials depend upon the type of material and the capacity of the testing machine available. For concrete, it is recommended that specimen length be equal to two diameters (L = 2D). For wood, rectangular specimens of $2 \times 2 \times 8''$, with long edges parallel to the grain, are recommended. Whenever possible, the L/D ratio should not be more than three lengths to one diameter. Fig. 8-11a illustrates a concrete specimen fracturing under compression.

Alignment of specimens on the testing machine is more critical in compression testing than in tensile testing. To increase uniaxial loading, the ends of the specimen should be perpendicular to its axis. To accomplish uniaxial loading and minimize friction, a spherical-end attachment is recommended for one end of the specimen. Fig. 8-11b. For specimens made from materials which can fracture or split easily at the ends, such as concrete or wood, it is desirable to cap the ends of the specimen to avoid chipping or splitting during testing.

Bending Tests

Many parts of structures (such as beams, columns, and floors) and sections of machines (such as shafts, machine bases, and chassis) are subjected to bending. To learn how various materials will behave in such circumstances, bending tests are conducted. Brittle and ductile materials such as metal, concrete, wood, plastics, brick, building stones, and gypsum, are subjected to bending tests. These tests, also called *transverse flexure tests,* are most frequently done to study such

Chapter 8. Testing of Industrial Materials

words, the notch helps to induce a "brittle" fracture in soft as well as hard materials and by doing so facilitates the study of impact behavior of materials. Therefore variations in the size and shape of the notch affect the impact behavior of the specimens. Loading velocities above a critical point will decrease the toughness value of certain materials. Low loading velocities, however, have no influence on the impact behavior of the specimen. Among the variables of the testing condition, the temperature of the specimen particularly affects the impact behavior. It

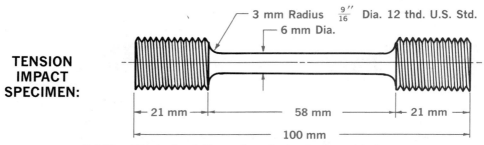

TENSION IMPACT SPECIMEN:

8-15b. Standard metallic specimen for tension impact test.

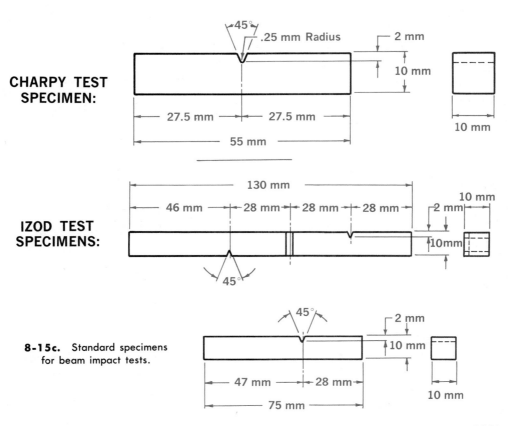

CHARPY TEST SPECIMEN:

IZOD TEST SPECIMENS:

8-15c. Standard specimens for beam impact tests.

Technology of Industrial Materials

has been observed that a decrease in temperature usually results in a relative decrease in toughness of the specimen.

Based on the way the specimen is loaded, impact tests are of three types: *torsion, tension,* and *beam*.

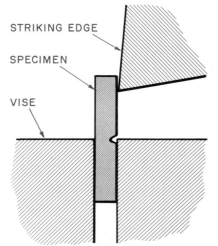

8-15d. Method of loading the Izod specimen.

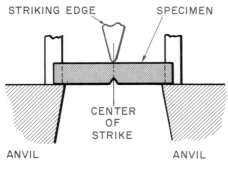

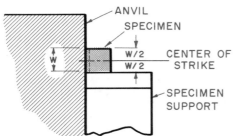

8-15e. Method of loading the Charpy specimen.

In torsion impact tests, the specimen is subjected to a combination of twisting and impact stresses. Torsion impact tests are rarely used.

In tension impact tests the specimen is subjected to a combination of tension and impact stresses. Tension impact tests are used to some extent but have not been standardized.

Beam impact tests, Fig. 8-15c, are the most common and are of two kinds, *Izod* and *Charpy,* which differ in the loading method. In the Izod test, the specimen is loaded as a *cantilever-beam,* and is struck on the notch side. Fig. 8-15d. In the Charpy test, the specimen is loaded as a *simple-beam* and is struck on the side opposite the notch. Fig. 8-15e. Both tests use notched specimens which are fractured in flexure. Izod and Charpy tests have been standardized by ASTM for metallic materials and plastics.

Various types of impact testing machines, Fig. 8-16, have been developed and are commercially available, but the universal impact testing machine is the most common. The major features of a universal single-blow pendulum-type impact machine are:

◆ The hammer mounted on the free end of the pendulum arm to provide the energy required to fracture the specimen.
◆ The anvil and the vise to support the specimen mounted on the base of the machine.
◆ The energy measuring device which indicates the energy absorbed by the specimen after fracture.

Universal impact testing machines are designed to fracture the specimen with a single blow and to indicate directly the energy absorbed by the specimen. This energy is the difference between the energy remaining in the hammer after the

Table 8-B

Comparative summary of basic characteristics of hardness tests.

Charact.	Brinell	Rockwell	Vickers	Microhardness	Scleroscope
Penetrators used	10mm. ball (5mm. for thin mats.)	120° Brale and 1/16, 1/8, 1/4 & 1/2 inch balls	136° Pyramid	136° Knoop	A small hammer dropped from a pre-determined height
Loads Applied	3000kg, 1500kg & 500kg	For regular—10kg minor load: 60kg, 100kg & 150kg major loads—For superficial —3kg minor; 15kg, 30kg & 45kg major	1gm to 120kg—most common 5kg to 30kg	1gm to 1kg	
Hardness Scales	One scale	For regular—15 scales are used, but B & C are the most common —For superficial—5 scales, but N & T are the most common	One scale	One scale	One scale
Hardness Numbers	Derived by dividing load by size of impression	Derived by dividing load by size of impression	Derived by dividing load by size of impression	Derived by dividing load by size of impression	Derived on the basis of rebound on heat-treated steel
Major Applications	Usually for large heavy parts such as structural steel, heavy castings, large forgings, etc.	Usually for small finished metallic parts such as sheet metal, wire, forming tools, cutting tools, gears, valves, etc.—Plastic parts or sheets	Mostly the same applications as Rockwell plus for highly finished parts and thin sections down to 0.005 in.	Usually for extremely thin and hard materials such as glasses, plated surfaces, coatings, foils, individual constituents and grains of mats.	Used for almost any application for small and large parts, for thin and thick parts, for metallic and non-metallic mats.
Surface Condition of Specimen	Relatively clean and flat	Clean and smooth surface required	Clean and smooth surface required	Very clean, smooth flat and often polished surface required	Clean and smooth surface required

Technology of Industrial Materials

A measuring microscope is usually used to measure the impression. Calibration of the machine can be accomplished by using standard hardness blocks or test specimens. Brinell hardness numbers depend on the applied load, the length of time the load is applied, and the size of the penetrating ball. The hardness numbers vary from 16 to 600 for 10mm ball and standard loads of 3000kg, 1500kg, and 500kg—the smaller the number, the softer the material. The ASTM recommends that for 3000kg load and 10mm ball the numbers should vary from 96 to 600; for 1500kg load, from 48 to 300; and for 500kg load, from 16 to 100. Table 8-C, pages 110–112. The load should be applied for 15 to 35 seconds and released.

Because of the large area covered by the impression, Brinell tests are particularly suited for relatively porous materials such as castings, forgings, and structural steel members. The tests are not recommended for extremely hard materials, for parts that have been surface-hardened (case-hardened), or for parts on which a large impression is objectionable. The tests are easy to perform but time consuming. In most cases, Brinell tests require that the specimen be clean and flat at the testing spot. To avoid bulging, the tests should *not* be performed close to the edge of the specimen; neither should the various impressions be made close to each other. Because of the deep penetration, Brinell tests give good indication of the internal conditions of the specimen. However, they should not be used with thin materials. The ASTM recommends that the minimum thickness of the specimen be at least 10 times the depth of the impression.

Rockwell Hardness Tests

Rockwell hardness tests are considered penetration tests because a penetrator of

8-23a. Regular bench-type Rockwell hardness testing machine.

specified size is forced by a known load into the surface of the specimen. Rockwell hardness numbers are derived on the basis of applied load and depth of penetration. Fig. 8-23. Rockwell hardness tests are probably the most widely used tests in this country. They are performed on testing machines which are operated manually or automatically. A typical Rockwell hardness testing machine consists of the load application, hardness indicating, and specimen supporting systems.

◆ The *load application system* consists of three major (standard) loads of 60kg, 100kg, and 150kg, and a minor load of 10kg applied before the major load.

◆ The *hardness indicating system* consists of a dial indicator (scale graduated from 0 to 100 hardness numbers), a minor load indicator, and the various penetrators. Table 8-C.

◆ The *specimen supporting system* consists of the adjustable screw, various types of anvils, the zero dial adjuster, and the de-

Chapter 8. Testing of Industrial Materials

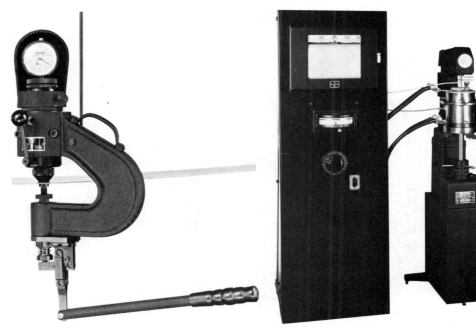

8-23b. Portable production-type Rockwell hardness testing machine.

8-23c. Rockwell hardness tester for elevated temperatures.

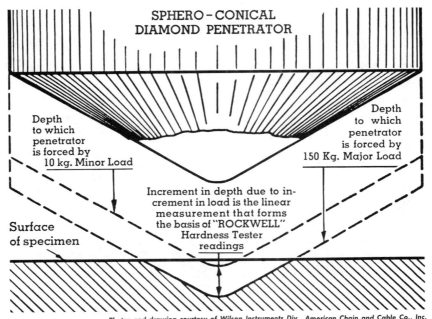

Photos and drawing courtesy of Wilson Instruments Div., American Chain and Cable Co., Inc.

8-23d. Penetration of major and minor loads in Rockwell hardness testing.

Table 8-C
Approximate hardness conversion numbers for steel.

Rockwell						1/16" Ball Penetrator	Vickers	Brinell	Micro-Hardness	Tensile Strength
Diamond Cone Penetrator "Brale"							136° Diamond Pyramid Penet.	10mm Steel Ball	Knoop Penetrator	Thousand lbs. per sq. in.
A	C	D	N	N	N	G				
60 kg	150 kg	100 kg	15 kg	30 kg	45 kg	150 kg	5–50 kg	3000 kg	500 gms	
86.5	70	78.5	94.0	86.0	77.5		1075		1029	
86.0	69	77.7	93.6	85.1	76.5		1007		988	
85.4	68	77.0	93.3	84.3	75.4		940		949	
85.0	67	76.2	92.9	83.5	74.3		900		912	
84.6	66	75.5	92.5	82.7	73.2		865		877	
84.1	65	74.7	92.1	81.8	72.1		832		844	
83.5	64	73.9	91.8	80.9	71.0		800		814	
82.9	63	73.1	91.5	80.0	69.9		772		787	
82.3	62	72.3	91.1	79.1	68.8		746		762	
81.7	61	71.5	90.7	78.2	67.7		720		738	
81.2	60	70.8	90.3	77.3	66.6		697		715	
80.6	59	70.0	89.9	76.4	65.5		674		693	
80.1	58	69.3	89.4	75.5	64.3		653		671	
79.6	57	68.5	89.0	74.6	63.1		633		650	
79.0	56	67.7	88.5	73.8	62.0		613		630	
78.5	55	67.0	88.0	72.9	60.8		595		611	301
78.0	54	66.2	87.5	72.0	59.6		577		593	292
77.5	53	65.5	87.0	71.1	58.4		560		575	283
77.0	52	64.7	86.5	70.2	57.2		544	500	558	272
76.4	51	63.9	86.0	69.3	56.0		528	487	542	264
75.9	50	63.1	85.5	68.4	54.9		513	475	527	256
75.3	49	62.3	85.0	67.5	53.7		498	464	513	246
74.8	48	61.5	84.5	66.6	52.5		484	451	500	237
74.3	47	60.8	84.0	65.7	51.4		471	442	487	229
73.7	46	60.0	83.5	64.8	50.2		458	432	475	221
73.2	45	59.3	83.0	63.9	49.0		446	421	463	215
72.7	44	58.5	82.4	63.0	47.8		434	409	451	208
72.1	43	57.7	81.9	62.1	46.7		423	400	440	201
71.6	42	56.9	81.4	61.2	45.5		412	390	429	194
71.0	41	56.2	80.9	60.3	44.3		402	381	419	188
70.5	40	55.4	80.4	59.5	43.1		392	371	409	182
70.0	39	54.6	79.8	58.6	41.9		382	362	400	177
69.5	38	53.8	79.3	57.7	40.7		372	353	391	171
69.0	37	53.1	78.8	56.8	39.6		363	344	383	166
68.5	36	52.3	78.3	55.9	38.4		354	336	375	162
68.0	35	51.5	77.7	55.0	37.2		345	327	367	157
67.5	34	50.8	77.1	54.1	36.0		336	319	359	153
67.0	33	50.0	76.6	52.1	34.9		327	311	351	149
66.5	32	49.3	76.0	52.2	33.7		318	301	343	145
66.0	31	48.5	75.5	51.3	32.5		310	294	335	142
65.5	30	47.7	75.0	50.4	31.4	92.0	302	286	328	139
65.0	29	47.0	74.4	49.5	30.2	91.0	294	279	322	135
64.5	28	46.3	73.9	48.6	29.0	90.0	286	271	316	132
64.0	27	45.5	73.3	47.7	27.8	89.0	279	264	310	129
63.5	26	44.7	72.7	46.8	26.6	88.0	272	258	305	126
63.0	25	43.9	72.1	45.9	25.5	86.9	266	253	300	124
62.5	24	43.2	71.6	45.0	24.3	85.8	260	247	295	121
62.0	23	42.5	71.1	44.1	23.1	84.6	254	243	291	119
61.5	22	41.7	70.5	43.2	21.9	83.4	248	237	287	116
61.0	21	40.9	70.0	42.3	20.7	82.2	243	231	283	114
60.5	20	40.1	69.5	41.4	19.6	81.0	238	226	280	112

Chapter 8. Testing of Industrial Materials

Table 8-C (Continued)
Approximate hardness conversion numbers for steel.

Rockwell					Sclero-scope	Mohs	Vickers 136° Diamond Pyramid Penet. 5–50 kg	Brinell 10mm Steel Ball 3000 kg	Micro-hardness Knoop Pene-trator 500 gms	Tensile Strength psi
B 100 kg	G 150 kg	T 15 kg	T 30 kg	T 45 kg						
100	82.5	93.0	82.0	72.0			234	240	280	
99	80.9	92.6	81.4	71.0			228	234	261	
98	79.2	92.3	80.8	70.1			222	228	246	
97	77.5	92.0	80.2	69.1	33		216	222	234	108
96	75.9	91.7	79.6	68.2	32		210	216	224	105
95	74.2	91.4	79.0	67.2	31		204	210	216	102
94	72.5	91.1	78.4	66.3	31		198	205	209	100
93	70.8	90.8	77.8	65.3	30		192	201	203	98
92	69.2	90.5	77.3	64.4	29		188	196	197	96
91	67.5	90.2	76.7	63.5	28	5.0	184	190	191	94
90	65.8	89.9	76.1	62.5	28		180	186	186	92
89	64.2	89.6	75.5	61.5	27		178	182	181	90
88	62.5	89.3	74.9	60.5	27		174	178	177	88
87	60.8	89.0	74.3	59.6	26		170	174	173	86
86	59.1	88.6	73.7	58.7	26		168	170	169	84
85	57.4	88.3	73.1	57.8	25		164	166	165	82
84	55.7	87.9	72.5	56.8	25		162	163	162	81
83	54.0	87.6	71.9	55.9	24		158	160	159	79
82	52.3	87.3	71.3	55.0	24		156	157	156	77
81	50.6	87.0	70.7	54.0	23		153	153	153	76
80	49.0	86.6	70.0	53.0	23	4.5	150	150	150	75
79	47.4	86.3	69.4	52.1	22		148	148	147	74
78	45.7	86.0	68.8	51.2	22		146	145	144	72
77	44.1	85.6	68.1	50.2			142	143	141	
76	42.5	85.3	67.5	49.3	21		140	140	138	71
75	40.8	85.0	66.8	48.4	21		138	138	136	70
74	39.2	84.6	66.2	47.5	21		135	136	134	68
73	37.5	84.3	65.5	46.5	20		132	133	132	66
72	35.9	84.0	64.8	45.5			130	131	129	
71	34.3	83.7	64.1	44.5			128	128	127	65
70	32.7	83.4	63.4	43.5			126	126	125	64
69	31.1	83.0	62.7	42.5			124	124	122	63
68	29.6	82.7	62.1	41.5			122	122	120	62
67	28.0	82.4	61.4	40.5			120	120	118	61
66	26.5	82.1	60.7	39.5			118	118	116	
65	25.0	81.8	60.0	38.5			116	116	114	60
64	23.5	81.5	59.3	37.4			114	114	112	59
63	22.0	81.1	58.6	36.4			112	112	110	58
62	20.5	80.8	57.9	35.4			110	110	108	56
61	19.0	80.5	57.2	34.4			108	108	107	
60	17.5	80.1	56.5	33.4			107	107	105	55
59	16.0	79.8	55.8	32.4			105	106	104	
58	14.5	79.5	55.2	31.3			104	104	102	54
57	13.0	79.2	54.5	30.3			102	103	102	53
56	11.5	78.8	53.8	29.3			100	101	99	
55	10.0	78.5	53.1	28.2			99	100	98	
54	8.5	78.2	52.4	27.2			98	99	97	
53	7.0	77.9	51.7	26.1			96	97	96	
52	5.5	77.5	51.0	25.1			95	96	94	
51	4.0	77.2	50.3	24.1			94	95	93	
50	2.5	76.9	49.7	23.1			93	94	92	

Table 8-C continued on following page.

Technology of Industrial Materials

Table 8-C (Continued)

Approximate hardness conversion numbers for steel.

Rockwell								Vickers	Brinell	Micro-Hardness
1/16" Ball Penetrator				1/8" Ball Penetrator			136° Diamond Pyramid Penetrator	10mm Steel Ball	Knoop Penetrator
B	F	T	T	T	E	H	K			
100kg	60kg	15kg	30kg	45kg	100kg	60kg	150kg	5–50kg	500kg	500gms
50	85.4	76.9	49.7	23.1	87.3		64.4	93	83.0	92
49	84.9	76.5	49.0	22.0	86.7		63.5	92	82.3	91
48	84.3	76.2	48.3	20.9	86.0		62.6	90	81.5	90
47	83.8	75.9	47.7	19.9	85.4		61.8	89	80.7	89
46	83.2	75.5	47.0	18.9	84.8		60.9	88	80.0	88
45	82.6	75.2	46.3	17.8	84.2		60.0	87	79.2	87
44	82.1	74.9	45.6	16.8	83.6		59.1	86	78.4	86
43	81.5	74.6	44.9	15.7	82.9		58.3	85	77.7	85
42	81.0	74.3	44.2	14.7	82.3		57.4	84	76.8	
41	80.4	74.0	43.5	13.6	81.7		56.5	83	76.0	84
40	79.8	73.7	42.9	12.5	81.0		55.6	82	75.2	83
39	79.2	73.4	42.2	11.5	80.4		54.8	81	74.5	
38	78.6	73.0	41.5	10.5	79.8		53.9		73.7	82
37	78.1	72.7	40.8	9.5	79.2		53.0	80	72.9	
36	77.5	72.4	40.2	8.4	78.6	100	52.2	79	72.0	81
35	76.9	72.1	39.5	7.4	78.0	99.7	51.3	78	71.1	
34	76.3	71.7	38.8	6.3	77.4	99.3	50.4		70.4	80
33	75.8	71.4	38.1	5.3	76.8	99.0	49.6	77	69.7	
32	75.2	71.1	37.4	4.2	76.2	98.6	48.8	76	68.9	
31	74.6	70.8	36.7	3.1	75.6	98.2	47.9	75	68.1	79
30	74.0	70.5	36.0	2.0	75.0	97.8	47.0		67.3	
29	73.4	70.1	35.3	1.0	74.4	97.5	46.1	74	66.7	78
28	72.9	69.8	34.6	0	73.8	97.2	45.3	73	66.0	77
27	72.3	69.5	33.9		73.2	96.8	44.4	72	65.3	
26	71.8	69.2	33.2		72.6	96.4	43.5		64.7	76
25	71.3	68.8	32.5		72.0	96.0	42.6	71	63.9	
24	70.7	68.5	31.8		71.4	95.7	41.8	70	63.2	75
23	70.1	68.2	31.2		70.7	95.3	40.9		62.6	
22	69.5	67.8	30.5		70.1	94.9	40.0	69	62.0	74
21	69.0	67.5	29.8		69.5	94.5	39.2	68	61.4	
20	68.5	67.2	29.0		68.9	94.1	38.3		60.6	73
19	67.9	66.8	28.3		68.3	93.8	37.5	67	60.2	
18	67.3	66.5	27.6		67.7	93.5	36.6	66	59.6	72
17	66.7	66.2	26.8		67.1	93.2	35.7	65	59.4	
16	66.2	65.9	26.1		66.5	92.8	34.8		59.0	71
15	65.6	65.5	25.4		65.9	92.4	34.0	64	58.6	
14	65.0	65.2	24.7		65.3	92.0	33.1	63	58.2	
13	64.4	64.9	24.0		64.7	91.6	32.3		58.0	70
12	63.8	64.6	23.3		64.1	91.3	31.4	62	57.6	
11	63.2	64.3	22.6		63.5	90.9	30.5	61	57.2	
10	62.6	64.0	21.9		62.8	90.5	29.6	60	56.8	
9	62.0	63.7	21.2		62.2	90.2	28.8		56.4	69
8	61.4	63.4	20.5		61.6	89.8	27.9	59	56.0	
7	60.9	63.0	19.8		61.0	89.5	27.0	58	55.8	
6	60.4	62.7	19.1		60.4	89.2	26.2		55.4	
5	59.9	62.3	18.4		59.8	88.8	25.3	57	55.0	68
4	59.3	62.0	17.7		59.2	88.4	24.4		54.8	
3	58.7	61.7	17.0		58.6	88.0	23.6	56	54.4	
2	58.1	61.3	16.3		58.0	87.7	22.7	55	54.0	
1	57.6	61.0	15.3		57.5	87.3	21.8		53.8	
0	57.1	60.7	14.8		57.0	87.0	21.0	54	53.4	67

Chapter 8. Testing of Industrial Materials

Table 8-D
Rockwell hardness scales, loads, penetrators and typical applications.

Scale	Major Load, kg	Dial Figures	Penetrator	Typical Applications
B	100	red	$\frac{1}{16}$-in. ball	Copper alloys, soft steels, aluminum alloys, malleable iron, etc.
C	150	black	diamond-Brale	Steel, hard cast irons, pearlitic malleable iron, titanium, deep case hardened steel, and other materials harder than B 100.
A	60	black	diamond-Brale	Cemented carbides, thin steel, and shallow case-hardened steel.
D	100	black	diamond-Brale	Thin steel and medium case hardened steel, and pearlitic malleable iron.
E	100	red	$\frac{1}{8}$-in. ball	Cast iron, aluminum and magnesium alloys, bearing metals.
F	60	red	$\frac{1}{16}$-in. ball	Annealed copper alloys, thin soft sheet metals.
G	150	red	$\frac{1}{16}$-in. ball	Malleable irons, copper-nickel-zinc and cupronickel alloys. Upper limit G 92 to avoid possible flattening of ball.
H	60	red	$\frac{1}{8}$-in. ball	Aluminum, zinc, lead.
K	150	red	$\frac{1}{8}$-in. ball	Bearing metals and other very soft or thin materials. Use smallest ball and heaviest load that does not give anvil effect.
L	60	red	$\frac{1}{4}$-in. ball	
M	100	red	$\frac{1}{4}$-in. ball	
P	150	red	$\frac{1}{4}$-in. ball	
R	60	red	$\frac{1}{2}$-in. ball	
S	100	red	$\frac{1}{2}$-in. ball	
V	150	red	$\frac{1}{2}$-in. ball	

pressor bar for releasing the major load. Fig. 8-23.

Rockwell testers have fifteen different scales, as explained in Table 8-D. Scales B and C are the most commonly used. Two types of penetrators are used:

◆ The diamond spheroconical penetrator (Brale) with an angle of 120°.
◆ Four steel ball penetrators of $\frac{1}{16}$, $\frac{1}{8}$, $\frac{1}{4}$ and $\frac{1}{2}$ inch diameters, the $\frac{1}{16}''$ being the most commonly used.

Some of the advantages of the Rockwell tests are that a wide range of hardness can be measured by using the different loads, scales, and penetrators. Also, the hardness numbers are read directly on the dial, eliminating possible errors in measurement. Employment of the minor load allows good contact of penetrator with specimen, thus eliminating errors in reading due to surface imperfections. Because a very small penetrator is used, tests can be performed on finished products.

To perform a Rockwell test accurately, the scale, penetrator, thickness of specimen, and type of anvil should be carefully selected. A guide for selecting the appro-

Technology of Industrial Materials

Table 8-E
Guide for selecting scales in terms of specimen thickness.

Thickness, in.	For 1/16 in. Ball Penetrator				
	Rockwell Superficial Scales			Regular Scales	
	15T	30T	45T	F	B
	Dial Reading	Dial Reading	Dial Reading	Dial Reading	Dial Reading
0.010	91	...	...	...	...
0.012	86	...	...	...	...
0.014	81	79	...	...	...
0.016	75	73	71	...	...
0.018	68	64	62	...	...
0.020	...	55	53	...	...
0.022	...	45	43	...	...
0.024	...	34	31	98	94
0.026	...	...	18	91	87
0.028	...	...	4	85	80
0.030	...	...	...	77	71
0.032	...	...	...	69	62
0.034	...	...	...	...	52
0.036	...	...	...	...	40
0.038	...	...	...	...	28
0.040	...	...	...	...	...

Thickness, in.	For Brale Penetrator				
	Rockwell Superficial Scales			Regular Scales	
	15N	30N	45N	A	C
	Dial Reading	Dial Reading	Dial Reading	Dial Reading	Dial Reading
0.006	92	...	...	...	...
0.008	90	...	...	...	...
0.010	88	...	...	...	...
0.012	83	82	77	...	...
0.014	76	78.5	74	...	...
0.016	68	74	72	86	...
0.018	...	66	68	84	...
0.020	...	57	63	82	...
0.022	...	47	58	79	69
0.024	...	...	51	76	67
0.026	...	...	37	71	65
0.028	...	...	20	67	62
0.030	...	...	...	60	57
0.032	...	...	...	...	52
0.034	...	...	...	...	45
0.036	...	...	...	...	37
0.038	...	...	...	...	28
0.040	...	...	...	...	20

Any greater thickness or hardness can be tested on the indicated scale.

ASTM

Chapter 8. Testing of Industrial Materials

Table 8-F
Rockwell superficial hardness scales.

Major Load, kg	Scales				
	N Scale, Diamond Penetrator	T Scale, $\frac{1}{16}$-in. Ball	W Scale, $\frac{1}{8}$-in. Ball	X Scale, $\frac{1}{4}$-in. Ball	Y Scale, $\frac{1}{2}$-in. Ball
15	15N	15T	15W	15X	15Y
30	30N	30T	30W	30X	30Y
45	45N	45T	45W	45X	45Y

ASTM

priate scale is given by ASTM in Table 8-E. The surface of the specimen should be flat or symmetrical, clean, and free from imperfections. To avoid bulging, impressions should *not* be made close to each other or close to the edge of the specimen. Specimens should be tested in single thickness and only from one side. It is recommended that at least three readings be taken and averaged for reliable results of any test. Before a test is made, the accuracy of the testing machine should be verified by employing standard testing blocks provided with the machine or by using materials of known hardness. Finally, to avoid confusion among various scales, Rockwell hardness numbers should be prefixed with the letter of the scale used—for example, $R_C 30$ or $R_B 65$.

Rockwell Superficial Test

The Rockwell superficial hardness test has been developed to test *hard-thin* materials, such as razor blades, which cannot be tested by regular tests described above. The same Rockwell tester mentioned earlier is employed for superficial tests, but different loads and scales are used. Three standard loads, 15kg, 30kg, and 45kg, and a 3kg minor load are used for the superficial test. Five scales have been standardized, but only two, N and T, are extensively used. Table 8-F.

Vickers Hardness Test

The Vickers hardness test is a penetration test employing a square pyramid penetrator, with an included angle of 136°, which is forced by a predetermined load into the surface of the specimen. The impression is measured optically using a microscope. The hardness numbers can be derived by dividing the load by the area of penetration. Table 8-C. The recommended formula is:

$$\text{Vickers Hardness Number (HV)} = \frac{\text{Load (kg)}}{\text{Area of impression (sq. mm)}}$$

Another formula is:

$$HV = \frac{1.8544P}{d^2}$$

In the second formula, P is the load in kg and d is the mean diagonal of the square pyramidal impression in mm.

Compared with other penetration tests, Vickers provides more accurate hardness measurements for small and large parts. The load can be varied from 1 gram to 120kg. The load is applied for 15 seconds and released automatically. For several reasons—small size of impressions, the use of diamond penetrators, varied loads, and precise measurement of the impression—Vickers tests can be used for hard and thin materials. The test is relatively

Technology of Industrial Materials

sensitive and is considered mostly as a surface test.

To increase accuracy, the specimen should be clean, with a smooth surface, and flat or symmetrical. This is especially true when light weights are used. Before starting a test, the accuracy of the machine should be verified by employing standard test blocks provided with the machine or by the use of materials of known hardness. Several readings should be taken and averaged in each test to increase reliability of results.

Microhardness Tests

Although microhardness tests have not been standardized in this country, they are growing in importance and application. Microhardness tests are basically penetration tests; the hardness numbers are derived by dividing the load by the area of impression. A diamond pyramid penetrator (Knoop) with an included angle of 136° is used. The *long diagonal* of the impression is optically measured with a microscope. Fig. 8-24.

Microhardness tests are considered laboratory tests and used primarily for research purposes. They can be used for testing extremely thin materials, such as coatings, films, and foils, and for testing individual grains or constituents in metallic materials. Specimens with flat, smooth, highly polished, and often etched surfaces are required. Special equipment is used for microhardness tests. Fig. 8-24. The load applied can vary from 1 gram to 1kg. A high-magnification microscope is attached to the machine for viewing the testing area and for measuring the impression.

Rebound (Dynamic) Hardness Tests

It has been pointed out that penetration tests are static tests, and the property they measure is resistance to penetration. The rebound test (such as the Shore Scleroscope) is a dynamic test—a hammer (load) is dropped from a fixed height onto the surface of the specimen, making a small impression. After the hammer hits the surface of the specimen, it rebounds to a portion of its original height because a certain amount of its energy has been used to make the small impression. The property measured is elastic resistance to penetration. Therefore, the hardness of a material under testing is proportional to the amount of hammer rebound—the higher the hammer rebounds, the harder the material.

Hardness numbers are determined on the basis of average rebound of the ham-

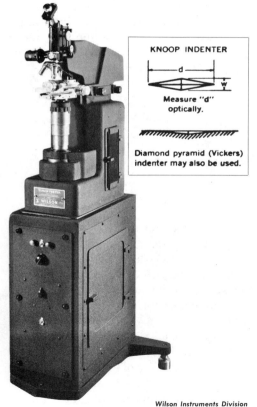

Wilson Instruments Division

8-24. Typical microhardness testing machine. Inset shows the geometry of the *Knoop* penetrator.

Chapter 8. Testing of Industrial Materials

mer on heat-treated carbon steel. The hardness scale is divided from 0 to 100 equal parts. Table 8-C. However, numbers higher than 100 can be used to test extremely hard materials.

The most representative hardness tester for rebound hardness is the *Shore Scleroscope*. Two models of these machines are commercially available—the vertical-scale model, Fig. 8-25a, and the dial-recording model, Fig. 8-25b. Both machines provide direct reading of rebound hardness and can be used to test small and large parts. Due to the light weight used, Shore tests can be employed for hard steel parts as thin as .005″. The machines can be bench mounted for laboratories or portable for field testing. Before testing, the machine should be checked for accuracy by using the reference bars of known hardness provided with the machine.

Scratch Hardness Tests

Two of the most common scratch hardness tests are the *file test* and the *Mohs scale*. The file test is an empirical test with no standard hardness scale. It is used for comparative hardness testing—that is, to find out whether one specimen is harder or softer than another. Although file tests are commonly done in many shops, considerable experience is required for accurate interpretation of results.

The Mohs scale is based on the known hardness of ten selected minerals. Numbers in sequence have been assigned to the ten minerals according to their hardness.

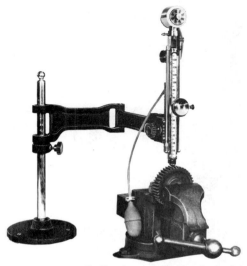

The Shore Instruments Manufacturing Co., Inc.
8-25a. Vertical-scale model Shore Scleroscope.

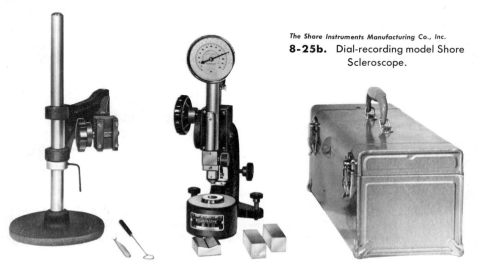

The Shore Instruments Manufacturing Co., Inc.
8-25b. Dial-recording model Shore Scleroscope.

Firestone Tire and Rubber Company

8-26a. This powerful scanning electron microscope utilizes a reflection technique in studying materials for research. The surface of the specimen is scanned with a fine electron beam. When the beam hits the specimen, electrons are given off and magnified, producing a detailed image of the material.

Talc, being the softest mineral, corresponds to number 1, and diamond, being the hardest mineral has the number 10. These ten numbers comprise the Mohs scale; specimens are tested and assigned numbers between 1 and 10 which indicate their hardness when compared with the known items on the scale. The scale is used primarily by mineralogists and geologists and has limited application in manufacturing.

TYPES OF TESTS—NONDESTRUCTIVE

Any test which does not destroy the specimen can be classified as nondestructive. Nondestructive tests have been used primarily for inspection purposes to locate defects in materials and products. However, due to recent developments explained in the following pages, nondestructive tests are increasingly employed to determine properties and characteristics of materials. Most nondestructive tests may be classified under five basic groups:

- Visual.
- Radiographic.
- Ultrasonic.
- Magnetic.
- Electrical.

The following discussion is centered on the basic concepts in each of these groups rather than on individual tests.

Visual Examination Tests

Visual examination tests are the oldest and most widely used inspection methods. They are used primarily to identify surface defects (such as cracks or porosity) induced by such manufacturing processes

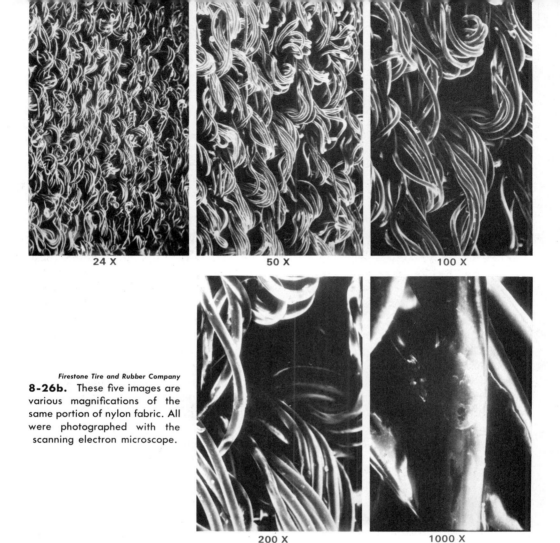

Firestone Tire and Rubber Company
8-26b. These five images are various magnifications of the same portion of nylon fabric. All were photographed with the scanning electron microscope.

as casting, welding, rolling, forging, machining, and heat-treating. Visual tests are inexpensive, simple, fast to perform, and very flexible. Some experience, however, is needed for efficient identification of defects and interpretation of results. The surface under inspection must be clean and well illuminated. Whenever possible such optical aids as microscopes or other magnifying lenses should be used. Fig. 8-26.

Liquid Penetrant Tests—Liquid penetrant tests are basically visual inspection tests. Highly visible (fluorescent or red) liquids are dipped, sprayed, or brushed over the area of the specimen to be examined. Through capillary action the liquid is pulled into the cavity of any surface defect, thus making the defect more visible.

One of the oldest liquid penetrant methods is the *oil-whiting* test employed to locate surface cracks that are otherwise difficult to detect. More recently, however, liquid penetrant tests have become available under such trade names as *Zyglo*,

Technology of Industrial Materials

Magnaflux Corporation

8-27. Typical surface defect located by liquid penetrant test.

The procedure for performing a liquid penetrant test is relatively simple. The surface of the specimen is cleaned and the liquid penetrant is applied. After the excess penetrant is removed, a developer in the form of a white powder suspension is applied and the defect is located and inspected. The function of the developer is to increase the visibility of the defect. Fig. 8-27.

Radiographic Tests

Industrial radiographic tests are basically photographic processes employing penetrating radiation instead of visible light. These methods are based on shortwave radiation which penetrates many solid materials. As radiation passes through the testing specimen, part of it is absorbed by the specimen. Most of the radiation passes through and leaves an image (photograph) on a film positioned behind the specimen. If a defective specimen is tested, a different amount of radiation will be absorbed by the defect from that absorbed by the rest of the specimen;

Spotcheck, Met-L-Chek, Dy-Chek, and *Dyeline.* These tests can be applied to any nonporous material such as plastics, metals and certain kinds of ceramics. However, the value of these tests is somewhat limited because only surface cracks can be located, and there is no means of determining the actual depth of the cracks.

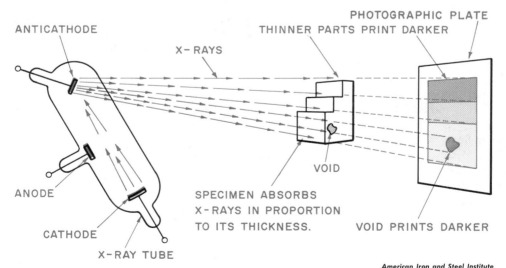

American Iron and Steel Institute

8-28. Schematic representation of the principle of radiographic testing.

Chapter 8. Testing of Industrial Materials

this will cause some areas of the film to be darker or lighter than surrounding areas, depending on the densities of the defect and of the specimen. Fig. 8-28.

The two most common sources of radiation employed in industrial radiography are *X-ray* and *gamma ray*. The main difference between the two is in the source of radiation. X-ray radiation is supplied by a tube housed in an X-ray machine. Gamma rays are supplied by a radioisotope generated in a nuclear reactor and housed in a gamma ray camera.

The machines for X-ray radiography are classified in terms of maximum voltage and are commercially available in various capacities from 15 to over 300 kilovolts. An X-ray machine consists of the X-ray tube which generates the radiation, the high voltage transformer, and the various operating controls. X-ray mathods are employed extensively in the welding and foundry industries. In welding, X-ray methods are considered among the most reliable for testing defective welds. The most common welding defects tested are inadequate penetration, porosity, slag inclusions, incomplete fusion, undercutting, and cracks. In the foundry industry, X-ray methods are used to test castings for such defects as inclusions, porosity, blowholes, cracks, and shrinkage defects. Fig. 8-29a.

Gamma rays are the result of disintegration of natural or man-made radioac-

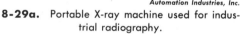

8-29a. Portable X-ray machine used for industrial radiography.

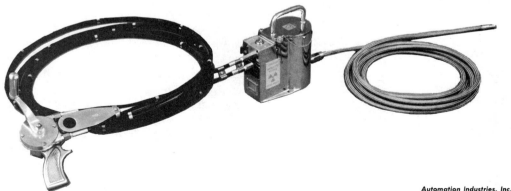

8-29b. Typical gamma ray camera used for industrial radiography.

8-30a. Ultrasonic unit employed in testing aerospace components.

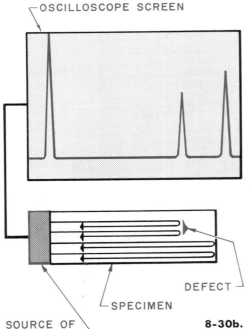

tive materials. Among the most important radioactive materials used are radium, iridium, and cobalt. Gamma ray cameras are small, portable, low cost, and can be used much more readily than X-ray machines. Fig. 8-29b, page 321.

One problem associated with gamma rays, and to lesser extent with X-rays, is safety. Like other materials, the human body is penetrated by and absorbs radiation; beyond a certain level, the amount absorbed becomes fatal. Among the various means developed for protection against radiation are shielding with lead, controlling the duration of exposure, and controlling the distance from the source. In addition to the basic radiographic methods discussed above, certain related methods have been developed in recent

8-30b. This drawing illustrates the principle of ultrasonic testing as explained in this text.

Chapter 8. Testing of Industrial Materials

years. Some of these methods are *fluoroscopy, xeroradiography,* and *electron-neutron radiography.*

Ultrasonic Tests

Sound waves (vibrational waves) with frequencies above the range of normal hearing are called *ultrasonic.* Properties of these waves which make them useful in testing are:

- They travel essentially in a straight line.
- They are transmittable by nearly all materials.

In testing, a beam of ultrasonic waves (vibrations) is generated by a source (usually a piezoelectric crystal) and sent through the specimen. If a defect (discontinuity) is present in the specimen, it will form a barrier to the ultrasonic waves and reflect some or all of them back to the source. The reflected waves are received by the testing instrument and converted into electric energy to measure the amplitude of the waves and the time of travel through the specimen. Some of the waves, however, are reflected from the opposite side of the specimen and are used as reference for the total length of the specimen. Fig. 8-30.

The purpose of ultrasonic tests is to determine the sizes, shapes, and locations of such surface and subsurface defects as cracks, inclusions, and blowholes. These tests are used to measure physical characteristics of materials, such as thickness, and to determine differences in the structure and properties of materials. Among the materials ultrasonically tested are metals, plastics, and ceramics.

Various types of ultrasonic testing instruments have been developed for particular applications. Ultrasonic testing instruments are employed extensively in automated high production lines for quality control and in laboratories for materials research. Lightweight, battery-operated ultrasonic instruments are used in the field for testing of structures and parts. Although the instruments are complicated electronic devices, the tests are simple to perform. However, some experience is required for interpreting the results.

Magnetic Analysis Tests

Magnetic analysis tests are based on the principle that the magnetic characteristics of a material are related to its composition and that changes in magnetic characteristics will therefore result from changes in composition. By observing the changes in the magnetic characteristics (magnetic field) of a material, the changes in its composition due to defects can be detected. Fig. 8-31. A magnetic flux (such as iron filings) is used for observing the changes in the magnetic field. When a

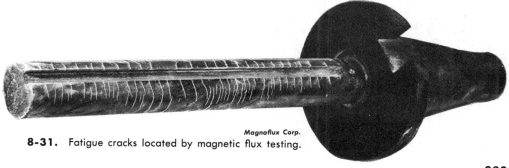

Magnaflux Corp.
8-31. Fatigue cracks located by magnetic flux testing.

323

Technology of Industrial Materials

specimen with no defects is magnetized and the magnetic flux is applied, the particles of the flux are not crowded at any particular spot but are dispersed along the path of the magnetic field. However, if the specimen has any such surface or subsurface defects as blowholes, cracks, and inclusions, the path of the magnetic flux is distorted because the defect has different composition and therefore different magnetic characteristics, than the rest of the specimen. In this case, some of the particles of the magnetic flux are crowded at the point of the defect because the defect acts as a *local magnet,* attracting and holding the magnetic flux at that spot.

Specimens can be magnetized in various ways. One method is by passing an electric current through them. This is referred to as *circular magnetization* and is good for locating *longitudinal defects.* Fig. 8-32.

A second method is by passing an electric current through a conductor which surrounds or is in contact with the specimen. This is referred to as *longitudinal magnetization* and is useful for locating defects which are *circumferential.* Fig. 8-32.

It is also possible to magnetize a part *multidirectionally.* This type of magnetization is recommended for extremely large parts such as large castings and forgings.

Still another method is to magnetize the specimen with magnets; however, this is not done extensively.

Most of the magnetic flux is made from black magnetic iron oxide ground into fine particles. (100 mesh sieve) and covered with fluorescent coatings to increase its visibility under black light. The flux can be applied by hand or automatically. It may be a dry powder (dry method); or it may be a water or oil suspension (wet method) applied by spraying or dipping.

Magnetic analysis tests are relatively easy to perform. They can be applied to magnetic specimens of various sizes and shapes. Four basic steps are involved in performing a test:

- Cleaning the surface.
- Magnetizing the specimen.
- Applying the magnetic flux.
- Inspecting for defects.

Various types of equipment are commercially available for magnetic analysis testing on a production line or in a laboratory. Portable units for field testing are also available. Fig. 8-33. Some of the advantages of magnetic analysis tests are:

- They can be used for locating surface and subsurface defects.

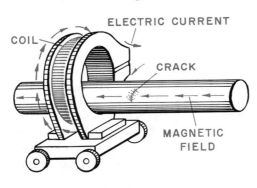
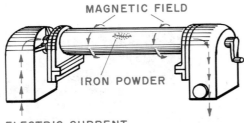

8-32. Schematic representation of the principles of circular and longitudinal magnetization.

residue undergoes cracking and coking reactions, and is deposited in the crucible. This residue is weighed and calculated as a percent of the original sample and reported as Conradson carbon residue. Fig. 8-38 illustrates the ASTM standard apparatus for determining Conradson carbon residue.

Cloud-Point and Pour-Point Tests

Small amounts of water, wax, and other impurities are contained in most petroleum products which can crystallize and separate out of the oil at certain temperatures. According to ASTM, the *cloud point* of an oil is the temperature at which a "cloud" of wax crystals forms at the bottom of the test jar when the oil is gradually cooled under prescribed conditions. The *pour point* of an oil is the lowest temperature at which the oil is observed to flow when gradually cooled under prescribed conditions.

The cloud point is important when the oil is exposed to low temperatures that may force the wax to crystallize, thus reducing the lubricating characteristics of the oil. The pour point indicates the lowest temperature at which the oil can flow, and therefore whether or not the oil can be used in gravity-fed lubricating systems that work below that temperature.

Both the cloud point and pour point tests have been standardized by ASTM and are performed in the same apparatus. Fig. 8-39. To determine the pour and cloud points, a sample is placed in the test jar, heated to about 120° F, gradually cooled at specified rates, and examined visually at intervals for flow characteristics or for wax crystals formed at the bottom of the jar.

Flash-Point and Fire-Point Tests

The *flash point* of a petroleum product is the temperature at which the oil vaporizes enough so that the vapors will ignite in the presence of a flame. The *fire point* is the temperature at which the oil, when heated but shielded from direct flame,

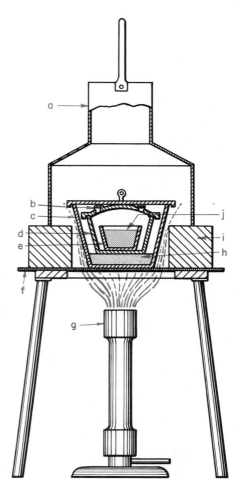

American Society for Testing and Materials
8-38. Standard apparatus for determining Conradson carbon residue in lubricating oils. (a) Hood. (b) Horizontal opening. (c) Outer iron crucible. (d) Inner iron crucible. (e) Porcelain crucible. (f) Wire support. (g) Meker type burner. (h) Dry sand. (i) Insulator. (j) Specimen.

Technology of Industrial Materials

ignites and continues to burn. Flash and fire points are indicators of evaporation and combustion of petroleum products; they are used to avoid possible fire hazards rather than as indicators of the lubricating characteristics of oils.

The flash point can be determined by three standard ASTM methods, depending on the material being tested. These methods are the *Tag-Closed Tester,* the *Pensky-Martens Closed Tester,* and the *Cleveland Open-Cup Method.*

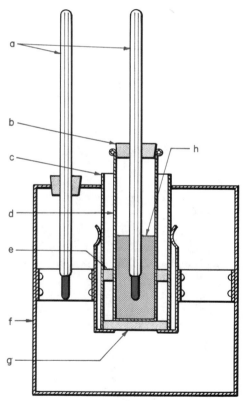

American Society for Testing and Materials

8-39. Standard apparatus for determining the cloud and pour points of lubricating oils. (a) Thermometer. (b) Cork. (c) Jacket. (d) Test jar. (e) Gasket. (f) Cooling bath. (g) Disk. (h) Oil for testing.

With the Cleveland Open-Cup Method (chosen for description here because it is the simplest method) a sample is placed in an open cup. It is then gradually heated to the temperature at which the sample will generate enough vapors to ignite (flash) at the surface of the cup in the presence of a flame. The temperature at the flash indicates the flash point of the oil. If the temperature of the same sample is raised still higher to the point at which the oil burns continuously, the temperature at this point indicates the fire point of the oil. Fig. 8-40 illustrates the ASTM standard apparatus used for determining the flash and fire points by the Cleveland Open-Cup Method.

Lubricating oils are also tested to determine the following:

◆ Sulfur content.

◆ Precipitation numbers (related to the separation of solid matter from the oil).

◆ Saponification numbers (related to the conversion of the oil into a substance chemically designated as a soap).

◆ Neutralization numbers (related to the degree of acidity or alkalinity).

◆ Such information as the effect of oils on engine wear (determined by mechanical tests).

Testing Gaseous Fuels

The two most important properties of gaseous fuels are *heating value* and *specific gravity.* The heating value test provides a measure of the heating energy of the gas as a fuel. The heating value test (given in *ASTM Standard* D900) is performed by burning 0.10 cu.ft. of gas in a standard water-flow type calorimeter.

The specific gravity test (given in *ASTM Standard* D1070) indicates the flow char-

APPENDIX A

Name, Atomic Symbol and Atomic Weight of Selected Chemical Elements

Name	Atomic Symbol	Atomic Weight	Name	Atomic Symbol	Atomic Weight
Aluminum	Al	26.97	Molybdenum	Mo	96.0
Antimony	Sb	121.76	Neodymium	Nd	144.27
Argon	A	39.944	Neon	Ne	20.183
Arsenic	As	74.91	Nickel	Ni	58.69
Barium	Ba	137.36	Nitrogen	N	14.008
Beryllium	Be	9.02	Osmium	Os	191.5
Bismuth	Bi	209.00	Oxygen	O	16.0000
Boron	B	10.82	Palladium	Pd	106.7
Bromine	Br	79.91	Phosphorus	P	31.02
Cadmium	Cd	112.41	Platinum	Pt	195.23
Calcium	Ca	40.08	Polonium	Po	—
Carbon	C	12.00	Potassium	K	39.096
Cerium	Ce	140.13	Praseodymium	Pr	140.92
Cesium	Cs	132.91	Protactinium	Pa	231
Chlorine	Cl	35.457	Radium	Ra	226.05
Chromium	Cr	52.01	Radon	Rn	222
Cobalt	Co	58.94	Rhenium	Re	186.31
Columbium	Cb	92.91	Rhodium	Rh	102.91
Copper	Cu	63.57	Rubidium	Rb	85.44
Dysprosium	Dy	162.46	Ruthenium	Ru	101.7
Erbium	Er	167.64	Samarium	Sm	150.43
Europium	Eu	152.0	Scandium	Sc	45.10
Fluorine	F	19.00	Selenium	Se	78.96
Gadolinium	Gd	157.30	Silicon	Si	28.06
Gallium	Ga	69.7	Silver	Ag	107.880
Germanium	Ge	72.62	Sodium	Na	22.997
Gold	Au	107.20	Strontium	Sr	87.63
Hafnium	Hf	178.6	Sulfur	S	32.06
Helium	He	4.002	Tantalum	Ta	180.88
Holmium	Ho	163.5	Tellurium	Te	127.61
Hydrogen	H	1.0078	Terbium	Tb	159.2
Illinium	Il	—	Thallium	Tl	204.39
Indium	In	114.76	Thorium	Th	232.12
Iodine	I	126.92	Thulium	Tm	169.4
Iridium	Ir	193.1	Tin	Sn	118.70
Iron	Fe	55.84	Titanium	Ti	47.90
Krypton	Kr	83.7	Tungsten	W	184.0
Lanthanum	La	138.92	Uranium	U	238.14
Lead	Pb	207.22	Vanadium	V	50.95
Lithium	Li	6.940	Xenon	Xe	131.3
Lutecium	Lu	175.0	Ytterbium	Yb	173.04
Magnesium	Mg	24.32	Yttrium	Y	88.92
Manganese	Mn	54.93	Zinc	Zn	65.38
Masurium	Ma	—	Zirconium	Zr	91.22
Mercury	Hg	200.61			

APPENDIX B

Typical Mechanical Properties of Wrought Aluminum Alloys[a]

Alloy and Temper	Tension Strength psi Ultimate	Tension Strength psi Yield	Elongation in 2 inches per cent 1/16-inch Thick Specimen	Elongation in 2 inches per cent 1/2-inch Diameter Specimen	Hardness Brinell Number 500-kg load 10-mm ball	Shear Shearing Strength psi	Fatigue Endurance[b] Limit psi	Modulus Modulus[c] of Elasticity psi
EC-O[d]	10,000	4,000	...	...[e]	...	8,000	.....	10.0×10^6
EC-H12	14,000	12,000	...	...	...	9,000	.....	10.0×10^6
EC-H14	16,000	14,000	...	...	...	10,000	.....	10.0×10^6
EC-H16	18,000	16,000	...	...	...	11,000	.....	10.0×10^6
EC-H18	21,000	19,000	...	...	...	.....	.....	10.0×10^6
EC-H19	27,000	24,000	...	...[f]	...	15,000	7,000	10.0×10^6
2EC-T6	32,000	28,000	...	19	70	22,000	9,000	10.0×10^6
2EC-T61	25,000	20,000	...	22	55	17,000	9,000	10.0×10^6
2EC-T62	30,000	25,000	...	20	65	21,000	9,000	10.0×10^6
2EC-T64	17,000	9,000	...	24	35	13,000	7,000	10.0×10^6
1060-O	10,000	4,000	43	...	19	7,000	3,000	10.0×10^6
1060-H12	12,000	11,000	16	...	23	8,000	4,000	10.0×10^6
1060-H14	14,000	13,000	12	...	26	9,000	5,000	10.0×10^6
1060-H16	16,000	15,000	8	...	30	10,000	6,500	10.0×10^6
1060-H18	19,000	18,000	6	...	35	11,000	6,500	10.0×10^6
1100-O	13,000	5,000	35	40	23	9,000	5,000	10.0×10^6
1100-H12	16,000	15,000	12	25	28	10,000	6,000	10.0×10^6
1100-H14	18,000	17,000	9	20	32	11,000	7,000	10.0×10^6
1100-H16	21,000	20,000	6	17	38	12,000	8,000	10.0×10^6
1100-H18	24,000	22,000	5	15	44	13,000	9,000	10.0×10^6
1345-O	12,000	4,000	40	...	22	8,000	.....	10.0×10^6
1345-H12	14,000	12,000	14	...	26	9,000	.....	10.0×10^6
1345-H14	16,000	14,000	11	...	30	10,000	.....	10.0×10^6
1345-H16	18,000	16,000	7	...	33	11,000	.....	10.0×10^6
1345-H18	21,000	19,000	6	...	38	12,000	.....	10.0×10^6
1345-H19	28,000	25,000	4	...	52	16,000	.....	10.0×10^6
2011-T3	55,000[g]	43,000[g]	...	15	95	32,000	18,000	10.2×10^6
2011-T6	57,000	39,000	...	17	97	34,000	18,000	10.2×10^6
2011-T8	59,000	45,000	...	12	100	35,000	18,000	10.2×10^6
2014-O	27,000	14,000	...	18	45	18,000	13,000	10.6×10^6
2014-T4,-T451	62,000	42,000[m]	...	20	105	38,000	20,000	10.6×10^6
2014-T6,-T651	70,000[h]	60,000[h]	...	13	135	42,000	18,000	10.6×10^6
Alclad 2014-O	25,000[i]	10,000	21	...	...	18,000	.....	10.5×10^6
Alclad 2014-T3	63,000[i]	40,000[i]	20	...	...	37,000	.....	10.5×10^6
Alclad 2014-T4,-T451	61,000[i]	37,000[i]	22	...	...	37,000	.....	10.5×10^6
Alclad 2014-T6,-T651	68,000[i]	60,000[i]	10	...	...	41,000	.....	10.5×10^6
2017-O	26,000	10,000	...	22	45	18,000	13,000	10.5×10^6
2017-T4,-T451	62,000	40,000	...	22	105	38,000	18,000	10.5×10^6
2018-T61	61,000	46,000	...	12	120	39,000	17,000	10.6×10^6
2024-O	27,000	11,000	20	22	47	18,000	13,000	10.6×10^6
2024-T3	70,000	50,000	18	...	120	41,000	20,000	10.6×10^6
2024-T36	72,000	57,000	13	...	130	42,000	18,000	10.6×10^6
2024-T4,-T351	68,000[h]	47,000[h]	20	19	120	41,000	20,000	10.6×10^6
2024-T6	69,000	57,000	...	10	125	41,000	18,000	10.6×10^6
2024-T81,-T851	70,000	65,000	6	...	128	43,000	18,000	10.6×10^6
2024-T86	75,000	71,000	6	...	135	45,000	18,000	10.6×10^6

APPENDIX B (Continued)
Typical Mechanical Properties of Wrought Aluminum Alloys[a]

Alloy and Temper	Tension Strength psi Ultimate	Tension Strength psi Yield	Elongation in 2 inches per cent 1/16-inch Thick Specimen	Elongation in 2 inches per cent 1/2-inch Diameter Specimen	Hardness Brinell Number 500-kg load 10-mm ball	Shear Shearing Strength psi	Fatigue Endurance[b] Limit psi	Modulus Modulus[c] of Elasticity psi
Alclad 2024-O	26,000	11,000	20	...	...	18,000	.....	10.6×10^6[o]
Alclad 2024-T3	65,000[j]	45,000[j]	18	...	...	40,000	.....	10.6×10^6[o]
Alclad 2024-T36	67,000[j]	53,000[j]	11	...	...	41,000	.....	10.6×10^6[o]
Alclad 2024-T4,-T351	64,000[j]	42,000[j]	19	...	...	40,000	.....	10.6×10^6[o]
Alclad 2024-T81,-T851	65,000[j]	60,000[j]	6	...	...	40,000	.....	10.6×10^6[o]
Alclad 2024-T86	70,000[j]	66,000[j]	6	...	...	42,000	.....	10.6×10^6[o]
2025-T6	58,000	37,000	...	19	110	35,000	18,000	10.4×10^6
2117-T4	43,000	24,000	...	27	70	28,000	14,000	10.3×10^6
2218-T72	48,000	37,000	...	11	95	30,000	.....	10.8×10^6
2219-O	25,000	10,000	20	...	...	.....	.....	10.6×10^6
2219-T31	54,000	37,000	17	...	100	33,000	.....	10.6×10^6
2219-T37	60,000	49,000	11	...	117	37,000	.....	10.6×10^6
2219-T62	61,000	42,000	11	...	115	37,000	15,000	10.6×10^6
2219-T81	70,000	53,000	11	...	130	41,000	15,000	10.6×10^6
2219-T87	70,000	58,000	10	...	130	41,000	15,000	10.6×10^6
3003-O	16,000	6,000	30	40	28	11,000	7,000	10.0×10^6
3003-H12	19,000	18,000	10	20	35	12,000	8,000	10.0×10^6
3003-H14	22,000	21,000	8	16	40	14,000	9,000	10.0×10^6
3003-H16	26,000	25,000	5	14	47	15,000	10,000	10.0×10^6
3003-H18	29,000	27,000	4	10	55	16,000	10,000	10.0×10^6
Alclad 3003-O	16,000	6,000	30	40	...	11,000	.....	10.0×10^6
Alclad 3003-H12	19,000	18,000	10	20	...	12,000	.....	10.0×10^6
Alclad 3003-H14	22,000	21,000	8	16	...	14,000	.....	10.0×10^6
Alclad 3003-H16	26,000	25,000	5	14	...	15,000	.....	10.0×10^6
Alclad 3003-H18	29,000	27,000	4	10	...	16,000	.....	10.0×10^6
3004-O	26,000	10,000	20	25	45	16,000	14,000	10.0×10^6
3004-H32	31,000	25,000	10	17	52	17,000	15,000	10.0×10^6
3004-H34	35,000	29,000	9	12	63	18,000	16,000	10.0×10^6
3004-H36	38,000	33,000	5	9	70	20,000	17,000	10.0×10^6
3004-H38	41,000	36,000	5	6	77	21,000	18,000	10.0×10^6
Alclad 3004-O	26,000	10,000	20	25	...	16,000	.....	10.0×10^6
Alclad 3004-H32	31,000	25,000	10	17	...	17,000	.....	10.0×10^6
Alclad 3004-H34	35,000	29,000	9	12	...	18,000	.....	10.0×10^6
Alclad 3004-H36	38,000	33,000	5	9	...	20,000	.....	10.0×10^6
Alclad 3004-H38	41,000	36,000	5	6	...	21,000	.....	10.0×10^6
3105-H25	26,000	24,000	8	...	47	16,000	.....	10.0×10^6
4032-T6	55,000	46,000	...	9	120	38,000	16,000	11.4×10^6
5005-O	18,000	6,000	30	...	30	11,000	.....	10.0×10^6
5005-H12	20,000	19,000	10	...	36	14,000	.....	10.0×10^6
5005-H14	23,000	22,000	6	...	41	14,000	.....	10.0×10^6
5005-H16	26,000	25,000	5	...	46	15,000	.....	10.0×10^6
5005-H18	29,000	28,000	4	...	51	16,000	.....	10.0×10^6
5005-H32	20,000	17,000	11	...	36	14,000	.....	10.0×10^6
5005-H34	23,000	20,000	8	...	41	14,000	.....	10.0×10^6
5005-H36	26,000	24,000	6	...	46	15,000	.....	10.0×10^6
5005-H38	29,000	27,000	5	...	51	16,000	.....	10.0×10^6
5050-O	21,000	8,000	24	...	36	15,000	13,000	10.0×10^6
5050-H32	25,000	21,000	9	...	46	17,000	15,000	10.0×10^6

APPENDIX B (Continued)
Typical Mechanical Properties of Wrought Aluminum Alloys[a]

Alloy and Temper	Tension Strength psi		Elongation in 2 inches per cent		Hardness Brinell Number 500-kg load 10-mm ball	Shear Shearing Strength psi	Fatigue Endurance[b] Limit psi	Modulus Modulus[c] of Elasticity psi
	Ultimate	Yield	1/16-inch Thick Specimen	1/2-inch Diameter Specimen				
5050-H34	28,000	24,000	8	...	53	18,000	16,000	10.0×10^6
5050-H36	30,000	26,000	7	...	58	19,000	17,000	10.0×10^6
5050-H38	32,000	29,000	6	...	63	20,000	18,000	10.0×10^6
5052-O	28,000	13,000	25	30	47	18,000	16,000	10.2×10^6
5052-H32	33,000	28,000	12	18	60	20,000	17,000	10.2×10^6
5052-H34	38,000	31,000	10	16	68	21,000	18,000	10.2×10^6
5052-H36	40,000	35,000	8	14	73	23,000	19,000	10.2×10^6
5052-H38	42,000	37,000	7	14	77	24,000	20,000	10.2×10^6
5056-O	42,000	22,000	...	35	65	26,000	20,000	10.3×10^6
5056-H18	63,000	59,000	...	10	105	34,000	22,000	10.3×10^6
5056-H38	60,000	50,000	...	15	100	32,000	22,000	10.3×10^6
5083-O	42,000	21,000	22	25	67	25,000	22,000	10.3×10^6
5083-H112[q]	44,000	28,000	16	...	80	26,000	22,000	10.3×10^6
5083-H113	46,000	33,000	16	16	82	28,000	23,000	10.3×10^6
5083-H323	47,000	36,000	10	...	84	27,000	.....	10.3×10^6
5083-H343	52,000	41,000	8	...	92	30,000	.....	10.3×10^6
5086-O	38,000	17,000	22	30	60	23,000	21,000	10.3×10^6
5086-H32	42,000	30,000	12	16	72	25,000	22,000	10.3×10^6
5086-H34	47,000	37,000	10	14	82	28,000	23,000	10.3×10^6
5086-H36	50,000	41,000	8	...	87	29,000	.....	10.3×10^6
5086-H112	39,000	19,000	14	...	64	23,000	21,000	10.3×10^6
5154-O	35,000	17,000	27	30	58	22,000	17,000	10.2×10^6
5154-H32	39,000	30,000	15	18	68	23,000	19,000	10.2×10^6
5154-H34	42,000	33,000	13	16	76	24,000	20,000	10.2×10^6
5154-H36	45,000	36,000	12	14	83	26,000	21,000	10.2×10^6
5154-H38	48,000	39,000	10	...	87	28,000	22,000	10.2×10^6
5154-H112	35,000	17,000	25	...	62	22,000	17,000	10.2×10^6
5254-O	35,000	17,000	27	30	58	22,000	17,000	10.2×10^6
5254-H32	39,000	30,000	15	18	68	23,000	19,000	10.2×10^6
5254-H34	42,000	33,000	13	16	76	24,000	20,000	10.2×10^6
5254-H36	45,000	36,000	12	14	83	26,000	21,000	10.2×10^6
5254-H38	48,000	39,000	10	...	87	28,000	22,000	10.2×10^6
5254-H112	35,000	17,000	25	...	62	22,000	17,000	10.2×10^6
5357-O	19,000	7,000	25	...	32	12,000	.....	10.0×10^6
5357-H25	27,000	23,000	10	...	50	16,000	.....	10.0×10^6
5357-H26	29,000	24,000	10	...	52	17,000	.....	10.0×10^6
5357-H38	32,000	30,000	6	...	55	18,000	.....	10.0×10^6
5454-O	36,000	17,000	22	25	60	23,000	19,000	10.2×10^6
5454-H32	40,000	30,000	10	18	73	24,000	20,000	10.2×10^6
5454-H34	44,000	35,000	10	16	81	26,000	21,000	10.2×10^6
5454-H112	36,000	18,000	20	...	60	23,000	.....	10.2×10^6
5454-H311	38,000	26,000	18	...	70	23,000	.....	10.2×10^6
5456-O	45,000	23,000	24	20	70	27,000	22,000	10.3×10^6
5456-H24	54,000	41,000	12	...	...	31,000	.....	10.3×10^6
5456-H112	45,000	24,000	22	...	70	27,000	.....	10.3×10^6
5456-H311	47,000	33,000	18	...	75	27,000	24,000	10.3×10^6
5456-H321	51,000	37,000	16	16	90	30,000	23,000	10.3×10^6
5456-H323	51,000	38,000	10	...	90	30,000	.....	10.3×10^6
5456-H343	56,000	43,000	8	...	94	33,000	.....	10.3×10^6

APPENDIX D (Continued)
The Metric System of Measurement
SI Unit Prefixes*

Multiplication Factor	Prefix	Symbol	Pronunciation (USA) (1)	Term (USA)	Term (Other Countries)
1 000 000 000 000 000 000 = 10^{18}	exa	E	as in Texas	one quintillion (2)	one trillion
1 000 000 000 000 000 = 10^{15}	peta	P	as in petal	one quadrillion (2)	one thousand billion
1 000 000 000 000 = 10^{12}	tera	T	as in terrace	one trillion (2)	one billion
1 000 000 000 = 10^{9}	giga	G	jig' a (a as in about)	one billion (2)	one milliard
1 000 000 = 10^{6}	mega	M	as in megaphone	one million	
1 000 = 10^{3}	kilo	k	as in kilowatt	one thousand	
100 = 10^{2}	hecto	h (3)	heck' toe	one hundred	
10 = 10	deka	da (3)	deck' a (a as in about)	ten	
0.1 = 10^{-1}	deci	d (3)	as in decimal	one tenth	
0.01 = 10^{-2}	centi	c (3)	as in sentiment	one hundredth	
0.001 = 10^{-3}	milli	m	as in military	one thousandth	
0.000 001 = 10^{-6}	micro	μ	as in microphone	one millionth	
0.000 000 001 = 10^{-9}	nano	n	nan' oh (nan as in Nancy)	one billionth (2)	one milliardth
0.000 000 000 001 = 10^{-12}	pico	p	peek' oh	one trillionth (2)	one billionth
0.000 000 000 000 001 = 10^{-15}	femto	f	fem' toe (fem as in feminine)	one quadrillionth (2)	one thousand billionth
0.000 000 000 000 000 001 = 10^{-18}	atto	a	as in anatomy	one quintillionth (2)	one trillionth

(1) The first syllable of every prefix is accented to assure that the prefix will retain its identity.
(2) These terms should be avoided in technical writing because the denominations above one million and below one millionth are different in most other countries, as indicated in the last column. Instead, use the prefixes or ten raised to an integral power.
(3) While hecto, deka, deci, and centi are SI prefixes, their use should generally be avoided except for the SI unit-multiples for area and volume and nontechnical use of centimeter, as for body and clothing measurement.

*Chart used by courtesy of the American National Metric Council.

APPENDIX D (Continued)
The Metric System of Measurement

SEVEN BASE UNITS

METRE — m LENGTH

The metre (common international spelling, metre) is defined as 1 650 763.73 wavelengths in vacuum of the orange-red line of the spectrum of krypton-86.

An interferometer is used to measure length by means of light waves.

The SI unit of area is the square metre (m^2).

The SI unit of volume is the cubic metre (m^3). The litre (0.001, cubic metre), although not an SI unit, is commonly used to measure fluid volume.

KILOGRAM — kg MASS

The standard for the unit of mass, the kilogram, is a cylinder of platinum-iridium alloy kept by the International Bureau of Weights and Measures at Paris. A duplicate in the custody of the National Bureau of Standards serves as the mass standard for the United States. This is the only base unit still defined by an artifact.

The SI unit of force is the newton (N). One newton is the force which, when applied to a 1 kilogram mass, will give the kilogram mass an acceleration of 1 (metre per second) per second.
$1N = 1kg \cdot m/s^2$

The SI unit for pressure is the pascal (Pa). $1Pa = 1N/m^2$

The SI unit for work and energy of any kind is the joule (J).
$1J = 1N \cdot m$

The SI unit for power of any kind is the watt (W). $1W = 1J/s$

SECOND — s TIME

The second is defined as the duration of 9 192 631 770 cycles of the radiation associated with a specified transition of the cesium-133 atom. It is realized by tuning an oscillator to the resonance frequency of cesium-133 atoms as they pass through a system of magnets and a resonant cavity into a detector.

Schematic diagram of an atomic beam spectrometer or "clock." Only those atoms whose magnetic moments are "flipped" in the transition region reach the detector. When 9 192 631 770 oscillations have occurred, the clock indicates one second has passed.

The number of periods or cycles per second is called frequency. The SI unit for frequency is the hertz (Hz). One hertz equals one cycle per second.

The SI unit for speed is the metre per second (m/s).

The SI unit for acceleration is the (metre per second) per second (m/s^2).

Standard frequencies and correct time are broadcast from WWV, WWVB, and WWVH, and stations of the U.S. Navy. Many shortwave receivers pick up WWV and WWVH on frequencies of 2.5, 5, 10, 15, and 20 megahertz.

AMPERE — A ELECTRIC CURRENT

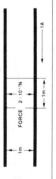

The ampere is defined as that current which, if maintained in each of two long parallel wires separated by one metre in free space, would produce a force between the two wires (due to their magnetic fields) of 2×10^{-7} newton for each metre of length.

The SI Unit of voltage is the volt (V).
$1V$ $1W/A$

The SI unit of electric resistance is the ohm (Ω).
1Ω $1V/A$

APPENDIX D (Continued)

Inches	Milli-metres	Inches	Milli-metres	Inches	Milli-metres	Inches	Milli-metres	Inches	Milli-metres	Inches	Milli-metres	Inches	Milli-metres	Inches	Milli-metres
0.50	12.700	1.50	38.100	2.50	63.500	3.50	88.900	4.50	114.300	5.50	139.700	6.50	165.100	7.50	190.500
.51	12.954	1.51	38.354	2.51	63.754	3.51	89.154	4.51	114.554	5.51	139.954	6.51	165.354	7.51	190.754
.52	13.208	1.52	38.608	2.52	64.008	3.52	89.408	4.52	114.808	5.52	140.208	6.52	165.608	7.52	191.008
.53	13.462	1.53	38.862	2.53	64.262	3.53	89.662	4.53	115.062	5.53	140.462	6.53	165.862	7.53	191.262
.54	13.716	1.54	39.116	2.54	64.516	3.54	89.916	4.54	115.316	5.54	140.716	6.54	166.116	7.54	191.516
.55	13.970	1.55	39.370	2.55	64.770	3.55	90.170	4.55	115.570	5.55	140.970	6.55	166.370	7.55	191.770
.56	14.224	1.56	39.624	2.56	65.024	3.56	90.424	4.56	115.824	5.56	141.224	6.56	166.624	7.56	192.024
.57	14.478	1.57	39.878	2.57	65.278	3.57	90.678	4.57	116.078	5.57	141.478	6.57	166.878	7.57	192.278
.58	14.732	1.58	40.132	2.58	65.532	3.58	90.932	4.58	116.332	5.58	141.732	6.58	167.132	7.58	192.532
.59	14.986	1.59	40.386	2.59	65.786	3.59	91.186	4.59	116.586	5.59	141.986	6.59	167.386	7.59	192.786
0.60	15.240	1.60	40.640	2.60	66.040	3.60	91.440	4.60	116.840	5.60	142.240	6.60	167.640	7.60	193.040
.61	15.494	1.61	40.894	2.61	66.294	3.61	91.694	4.61	117.094	5.61	142.494	6.61	167.894	7.61	193.294
.62	15.748	1.62	41.148	2.62	66.548	3.62	91.948	4.62	117.348	5.62	142.748	6.62	168.148	7.62	193.548
.63	16.002	1.63	41.402	2.63	66.802	3.63	92.202	4.63	117.602	5.63	143.002	6.63	168.402	7.63	193.802
.64	16.256	1.64	41.656	2.64	67.050	3.64	92.456	4.64	117.856	5.64	143.256	6.64	168.656	7.64	194.056
.65	16.510	1.65	41.910	2.65	67.310	3.65	92.710	4.65	118.110	5.65	143.510	6.65	168.910	7.65	194.316
.66	16.764	1.66	42.164	2.66	67.564	3.66	92.964	4.66	118.364	5.66	143.764	6.66	169.164	7.66	194.564
.67	17.018	1.67	42.418	2.67	67.818	3.67	93.218	4.67	118.618	5.67	144.018	6.67	169.418	7.67	194.818
.68	17.272	1.68	42.672	2.68	68.072	3.68	93.472	4.68	118.872	5.68	144.272	6.68	169.672	7.68	195.072
.69	17.526	1.69	42.926	2.69	68.326	3.69	93.726	4.69	119.126	5.69	144.526	6.69	169.926	7.69	195.326
0.70	17.780	1.70	43.180	2.70	68.580	3.70	93.980	4.70	119.380	5.70	144.780	6.70	170.180	7.70	195.580
.71	18.034	1.71	43.434	2.71	68.834	3.71	94.234	4.71	119.634	5.71	145.034	6.71	170.434	7.71	195.834
.72	18.288	1.72	43.688	2.72	69.088	3.72	94.488	4.72	119.888	5.72	145.288	6.72	170.688	7.72	196.088
.73	18.542	1.73	43.942	2.73	69.342	3.73	94.742	4.73	120.142	5.73	145.542	6.73	170.942	7.73	196.342
.74	18.796	1.74	44.196	2.74	69.596	3.74	94.996	4.74	120.396	5.74	145.796	6.74	171.196	7.74	196.596
.75	19.050	1.75	44.450	2.75	69.850	3.75	95.250	4.75	120.650	5.75	146.050	6.75	171.450	7.75	196.850
.76	19.304	1.76	44.704	2.76	70.104	3.76	95.504	4.76	120.904	5.76	146.304	6.76	171.704	7.76	197.104
.77	19.558	1.77	44.958	2.77	70.358	3.77	95.758	4.77	121.158	5.77	146.558	6.77	171.958	7.77	197.358
.78	19.812	1.78	45.212	2.78	70.612	3.78	96.012	4.78	121.412	5.78	146.812	6.78	172.212	7.78	197.612
.79	20.066	1.79	45.466	2.79	70.866	3.79	96.266	4.79	121.666	5.79	147.066	6.79	172.466	7.79	197.866
0.80	20.320	1.80	45.720	2.80	71.120	3.80	96.520	4.80	121.920	5.80	147.320	6.80	172.720	7.80	198.120
.81	20.574	1.81	45.974	2.81	71.374	3.81	96.774	4.81	122.174	5.81	147.574	6.81	172.974	7.81	198.374
.82	20.828	1.82	46.228	2.82	71.628	3.82	97.028	4.82	122.428	5.82	147.828	6.82	173.228	7.82	198.628
.83	21.082	1.83	46.482	2.83	71.882	3.83	97.282	4.83	122.682	5.83	148.082	6.83	173.482	7.83	198.882
.84	21.336	1.84	46.736	2.84	72.136	3.84	97.536	4.84	122.936	5.84	148.336	6.84	173.736	7.84	199.136
.85	21.590	1.85	46.990	2.85	72.390	3.85	97.790	4.85	123.190	5.85	148.590	6.85	173.990	7.85	199.390
.86	21.844	1.86	47.244	2.86	72.644	3.86	98.044	4.86	123.444	5.86	148.844	6.86	174.244	7.86	199.644
.87	22.098	1.87	47.498	2.87	72.898	3.87	98.298	4.87	123.698	5.87	149.098	6.87	174.498	7.87	199.898
.88	22.352	1.88	47.752	2.88	73.152	3.88	98.552	4.88	123.952	5.88	149.352	6.88	174.752	7.88	200.152
.89	22.606	1.89	48.006	2.89	73.406	3.89	98.806	4.89	124.206	5.89	149.606	6.89	175.006	7.89	200.406
0.90	22.860	1.90	48.260	2.90	73.660	3.90	99.060	4.90	124.460	5.90	149.860	6.90	175.260	7.90	200.660
.91	23.114	1.91	48.514	2.91	73.914	3.91	99.314	4.91	124.714	5.91	150.114	6.91	175.514	7.91	200.914
.92	23.368	1.92	48.768	2.92	74.168	3.92	99.568	4.92	124.968	5.92	150.368	6.92	175.768	7.92	201.168
.93	23.622	1.93	49.022	2.93	74.422	3.93	99.822	4.93	125.222	5.93	150.622	6.93	176.022	7.93	201.422
.94	23.876	1.94	49.276	2.94	74.676	3.94	100.076	4.94	125.476	5.94	150.876	6.94	176.276	7.94	201.676
.95	24.130	1.95	49.530	2.95	74.930	3.95	100.330	4.95	125.730	5.95	151.130	6.95	176.530	7.95	201.930
.96	24.384	1.96	49.784	2.96	75.184	3.96	100.584	4.96	125.984	5.96	151.384	6.96	176.784	7.96	202.184
.97	24.638	1.97	50.038	2.97	75.438	3.97	100.838	4.97	126.238	5.97	151.638	6.97	177.038	7.97	202.438
.98	24.892	1.98	50.292	2.98	75.692	3.98	101.092	4.98	126.492	5.98	151.892	6.98	177.292	7.98	202.692
.99	25.146	1.99	50.546	2.99	75.946	3.99	101.346	4.99	126.746	5.99	152.146	6.99	177.546	7.99	202.946

Inches	Milli-metres	Inches	Milli-metres
8.50	215.900	9.50	241.300
8.51	216.154	9.51	241.554
8.52	216.408	9.52	241.808
8.53	216.662	9.53	242.062
8.54	216.916	9.54	242.316
8.55	217.170	9.55	242.570
8.56	217.424	9.56	242.824
8.57	217.678	9.57	243.078
8.58	217.932	9.58	243.332
8.59	218.186	9.59	243.586
8.60	218.440	9.60	243.840
8.61	218.694	9.61	244.094
8.62	218.948	9.62	244.348
8.63	219.202	9.63	244.602
8.64	219.456	9.64	244.856
8.65	219.710	9.65	245.110
8.66	219.964	9.66	245.364
8.67	220.218	9.67	245.618
8.68	220.472	9.68	245.872
8.69	220.726	9.69	246.126
8.70	220.980	9.70	246.380
8.71	221.234	9.71	246.634
8.72	221.488	9.72	246.888
8.73	221.742	9.73	247.142
8.74	221.996	9.74	247.396
8.75	222.250	9.75	247.650
8.76	222.504	9.76	247.904
8.77	222.758	9.77	248.158
8.78	223.012	9.78	248.412
8.79	223.266	9.79	248.666
8.80	223.520	9.80	248.920
8.81	223.774	9.81	249.174
8.82	224.028	9.82	249.428
8.83	224.282	9.83	249.682
8.84	224.536	9.84	249.936
8.85	224.790	9.85	250.190
8.86	225.044	9.86	250.444
8.87	225.298	9.87	250.698
8.88	225.552	9.88	250.952
8.89	225.806	9.89	251.206
8.90	226.060	9.90	251.460
8.91	226.314	9.91	251.714
8.92	226.568	9.92	251.968
8.93	226.822	9.93	252.222
8.94	227.076	9.94	252.476
8.95	227.330	9.95	252.730
8.96	227.584	9.96	252.984
8.97	227.838	9.97	253.238
8.98	228.092	9.98	253.492
8.99	228.346	9.99	253.746

*ASTM

APPENDIX D (Continued)
Temperatures—Celsius to Fahrenheit
Conversion Table for Temperatures above 0° C

Temp. °C	0	1	2	3	4	5	6	7	8	9
0	32.0	33.8	35.6	37.4	39.2	41.0	42.8	44.6	46.4	48.2
10	50.0	51.8	53.6	55.4	57.2	59.0	60.8	62.6	64.4	66.2
20	68.0	69.8	71.6	73.4	75.2	77.0	78.8	80.6	82.4	84.2
30	86.0	87.8	89.6	91.4	93.2	95.0	96.8	98.6	100.4	102.2
40	104.0	105.8	107.6	109.4	111.2	113.0	114.8	116.6	118.4	120.2
50	122.0	123.8	125.6	127.4	129.2	131.0	132.8	134.6	136.4	138.2
60	140.0	141.8	143.6	145.4	147.2	149.0	150.8	152.6	154.4	156.2
70	158.0	159.8	161.6	163.4	165.2	167.0	168.8	170.6	172.4	174.2
80	176.0	177.8	179.6	181.4	183.2	185.0	186.8	188.6	190.4	192.2
90	194.0	195.8	197.6	199.4	201.2	203.0	204.8	206.6	208.4	210.2
100	212.0	213.8	215.6	217.4	219.2	221.0	222.8	224.6	226.4	228.2
110	230.0	231.8	233.6	235.4	237.2	239.0	240.8	242.6	244.4	246.2
120	248.0	249.8	251.6	253.4	255.2	257.0	258.8	260.6	262.4	264.2
130	266.0	267.8	269.6	271.4	273.2	275.0	276.8	278.6	280.4	282.2
140	284.0	285.8	287.6	289.4	291.2	293.0	294.8	296.6	298.4	300.2
150	302.0	303.8	305.6	307.4	309.2	311.0	312.8	314.6	316.4	318.2
160	320.0	321.8	323.6	325.4	327.2	329.0	330.8	332.6	334.4	336.2
170	338.0	339.8	341.6	343.4	345.2	347.0	348.8	350.6	352.4	354.2
180	356.0	357.8	359.6	361.4	363.2	365.0	366.8	368.6	370.4	372.2
190	374.0	375.8	377.6	379.4	381.2	383.0	384.8	386.6	388.4	390.2
200	392.0	393.8	395.6	397.4	399.2	401.0	402.8	404.6	406.4	408.2
210	410.0	411.8	413.6	415.4	417.2	419.0	420.8	422.6	424.4	426.2
220	428.0	429.8	431.6	433.4	435.2	437.0	438.8	440.6	442.4	444.2
230	446.0	447.8	449.6	451.4	453.2	455.0	456.8	458.6	460.4	462.2
240	464.0	465.8	467.6	469.4	471.2	473.0	474.8	476.6	478.4	480.2
250	482.0	483.8	485.6	487.4	489.2	491.0	492.8	494.6	496.4	498.2
260	500.0	501.8	503.6	505.4	507.2	509.0	510.8	512.6	514.4	516.2
270	518.0	519.8	521.6	523.4	525.2	527.0	528.8	530.6	532.4	534.2
280	536.0	537.8	539.6	541.4	543.2	545.0	546.8	548.6	550.4	552.2
290	554.0	555.8	557.6	559.4	561.2	563.0	564.8	566.6	568.4	570.2
300	572.0	573.8	575.6	577.4	579.2	581.0	582.8	584.6	586.4	588.2
310	590.0	591.8	593.6	595.4	597.2	599.0	600.8	602.6	604.4	606.2
320	608.0	609.8	611.6	613.4	615.2	617.0	618.8	620.6	622.4	624.2
330	626.0	627.8	629.6	631.4	633.2	635.0	636.8	638.6	640.4	642.2
340	644.0	645.8	647.6	649.4	651.2	653.0	654.8	656.6	658.4	660.2
350	662.0	663.8	665.6	667.4	669.2	671.0	672.8	674.6	676.4	678.2
360	680.0	681.8	683.6	685.4	687.2	689.0	690.8	692.6	694.4	696.2
370	698.0	699.8	701.6	703.4	705.2	707.0	708.8	710.6	712.4	714.2
380	716.0	717.8	719.6	721.4	723.2	725.0	726.8	728.6	730.4	732.2
390	734.0	735.8	737.6	739.4	741.2	743.0	744.8	746.6	748.4	750.2
400	752.0	753.8	755.6	757.4	759.2	761.0	762.8	764.6	766.4	768.2
410	770.0	771.8	773.6	775.4	777.2	779.0	780.8	782.6	784.4	786.2
420	788.0	789.8	791.6	793.4	795.2	797.0	798.8	800.6	802.4	804.2
430	806.0	807.8	809.6	811.4	813.2	815.0	816.8	818.6	820.4	822.2
440	824.0	825.8	827.6	829.4	831.2	833.0	834.8	836.6	838.4	840.2
450	842.0	843.8	845.6	847.4	849.2	851.0	852.8	854.6	856.4	858.2
460	860.0	861.8	863.6	865.4	867.2	869.0	870.8	872.6	874.4	876.2
470	878.0	879.8	881.6	883.4	885.2	887.0	888.8	890.6	892.4	894.2
480	896.0	897.8	899.6	901.4	903.2	905.0	906.8	908.6	910.4	912.2
490	914.0	915.8	917.6	919.4	921.2	923.0	924.8	926.6	928.4	930.2

APPENDIX D (Continued)
Temperatures—Celsius to Fahrenheit
Conversion Table for Temperatures above 0° C

Temp. °C.	0	1	2	3	4	5	6	7	8	9
500	932.0	933.8	935.6	937.4	939.2	941.0	942.8	944.6	946.4	948.2
510	950.0	951.8	953.6	955.4	957.2	959.0	960.8	962.6	964.4	966.2
520	968.0	969.8	971.6	973.4	975.2	977.0	978.8	980.6	982.4	984.2
530	986.0	987.8	989.6	991.4	993.2	995.0	996.8	998.6	1000.4	1002.2
540	1004.0	1005.8	1007.6	1009.4	1011.2	1013.0	1014.8	1016.6	1018.3	1020.2
550	1022.0	1023.8	1025.6	1027.4	1029.2	1031.0	1032.8	1034.6	1036.4	1038.2
560	1040.0	1041.8	1043.6	1045.4	1047.2	1049.0	1050.8	1052.6	1054.4	1056.2
570	1058.0	1059.8	1061.6	1063.4	1065.2	1067.0	1068.8	1070.6	1072.4	1074.2
580	1076.0	1077.8	1079.6	1081.4	1083.2	1085.0	1086.8	1088.6	1090.4	1092.2
590	1094.0	1095.8	1097.6	1099.4	1101.2	1103.0	1104.8	1106.6	1108.4	1110.2
600	1112.0	1113.8	1115.6	1117.4	1119.2	1121.0	1122.8	1124.6	1126.4	1128.2
610	1130.0	1131.8	1133.6	1135.4	1137.2	1139.0	1140.8	1142.6	1144.4	1146.2
620	1148.0	1149.8	1151.6	1153.4	1155.2	1157.0	1158.8	1160.6	1162.4	1164.2
630	1166.0	1167.8	1169.6	1171.4	1173.2	1175.0	1176.8	1178.6	1180.4	1182.2
640	1184.0	1185.8	1187.6	1189.4	1191.2	1193.0	1194.8	1196.6	1198.4	1200.2
650	1202.0	1203.8	1205.6	1207.4	1209.2	1211.0	1212.8	1214.6	1216.4	1218.2
660	1220.0	1221.8	1223.6	1225.4	1227.2	1229.0	1230.8	1232.6	1234.4	1236.2
670	1238.0	1239.8	1241.6	1243.4	1245.2	1247.0	1248.8	1250.6	1252.4	1254.2
680	1256.0	1257.8	1259.6	1261.4	1263.2	1265.0	1266.8	1268.6	1270.4	1272.2
690	1274.0	1275.8	1277.6	1279.4	1281.2	1283.0	1284.8	1286.6	1288.4	1290.2
700	1292.0	1293.8	1295.6	1297.4	1299.2	1301.0	1302.8	1304.6	1306.4	1308.2
710	1310.0	1311.8	1313.6	1315.4	1317.2	1319.0	1320.8	1322.6	1324.4	1326.2
720	1328.0	1329.8	1331.6	1333.4	1335.2	1337.0	1338.8	1340.6	1342.4	1344.2
730	1346.0	1347.8	1349.6	1351.4	1353.2	1355.0	1356.8	1358.6	1360.4	1362.2
740	1364.0	1365.8	1367.6	1369.4	1371.2	1373.0	1374.8	1376.6	1378.4	1380.2
750	1382.0	1383.8	1385.6	1387.4	1389.2	1391.0	1392.8	1394.6	1396.4	1398.2
760	1400.0	1401.8	1403.6	1405.4	1407.2	1409.0	1410.8	1412.6	1414.4	1416.2
770	1418.0	1419.8	1421.6	1423.4	1425.2	1427.0	1428.8	1430.6	1432.4	1434.2
780	1436.0	1437.8	1439.6	1441.4	1443.2	1445.0	1446.8	1448.6	1450.4	1452.2
790	1454.0	1455.8	1457.6	1459.4	1461.2	1463.0	1464.8	1466.6	1468.4	1470.2
800	1472.0	1473.8	1475.6	1477.4	1479.2	1481.0	1482.8	1484.6	1486.4	1488.2
810	1490.0	1491.8	1493.6	1495.4	1497.2	1499.0	1500.8	1502.6	1504.4	1506.2
820	1508.0	1509.8	1511.6	1513.4	1515.2	1517.0	1518.8	1520.6	1522.4	1524.2
830	1526.0	1527.8	1529.6	1531.4	1533.2	1535.0	1536.8	1538.6	1540.4	1542.2
840	1544.0	1545.8	1547.6	1549.4	1551.2	1553.0	1554.8	1556.6	1558.4	1560.2
850	1562.0	1563.8	1565.6	1567.4	1569.2	1571.0	1572.8	1574.6	1576.4	1578.2
860	1580.0	1581.8	1583.6	1585.4	1587.2	1589.0	1590.8	1592.6	1594.4	1596.2
870	1598.0	1599.8	1601.6	1603.4	1605.2	1607.0	1608.8	1610.6	1612.4	1614.2
880	1616.0	1617.8	1619.6	1621.4	1623.2	1625.0	1626.8	1628.8	1630.4	1632.2
890	1634.0	1635.8	1637.6	1639.4	1641.2	1643.0	1644.8	1646.6	1648.4	1650.2
900	1652.0	1653.8	1655.6	1657.4	1659.2	1661.0	1662.8	1664.6	1666.4	1668.2
910	1670.0	1671.8	1673.6	1675.4	1677.2	1679.0	1680.8	1682.6	1684.4	1686.2
920	1688.0	1689.8	1691.6	1693.4	1695.2	1697.0	1698.8	1700.6	1702.4	1704.2
930	1706.0	1707.8	1709.6	1711.4	1713.2	1715.0	1716.8	1718.6	1720.4	1722.2
940	1724.0	1725.8	1727.6	1729.4	1731.2	1733.0	1734.8	1736.6	1738.4	1740.2
950	1742.0	1743.8	1745.6	1747.4	1749.2	1751.0	1752.8	1754.6	1756.4	1758.2
960	1760.0	1761.8	1763.6	1765.4	1767.2	1769.0	1770.8	1772.6	1774.4	1776.2
970	1778.0	1779.8	1781.6	1783.4	1785.2	1787.0	1788.8	1790.6	1792.4	1794.2
980	1796.0	1797.8	1799.6	1801.4	1803.2	1805.0	1806.8	1808.6	1810.4	1812.2
990	1814.0	1815.8	1817.6	1819.4	1821.2	1823.0	1824.8	1826.6	1828.4	1830.2

APPENDIX D (Continued)
Temperatures—Celsius to Fahrenheit
Conversion Table for Temperatures above 0° C

Temp. °C	0	1	2	3	4	5	6	7	8	9
1000	1832.0	1833.8	1835.6	1837.4	1839.2	1841.0	1842.8	1844.6	1846.4	1848.2
1010	1850.0	1851.8	1853.6	1855.4	1857.2	1859.0	1860.8	1862.6	1864.4	1866.2
1020	1868.0	1869.8	1871.6	1873.4	1875.2	1877.0	1878.8	1880.6	1882.4	1884.2
1030	1886.0	1887.8	1889.6	1891.4	1893.2	1895.0	1896.8	1898.6	1900.4	1902.2
1040	1904.0	1905.8	1907.6	1909.4	1911.2	1913.0	1914.8	1916.6	1918.4	1920.2
1050	1922.0	1923.8	1925.6	1927.4	1929.2	1931.0	1932.8	1934.6	1936.4	1938.2
1060	1940.0	1941.8	1943.6	1945.4	1947.2	1949.0	1950.8	1952.6	1954.4	1956.2
1070	1958.0	1959.8	1961.6	1963.4	1965.2	1967.0	1968.8	1970.6	1972.4	1974.2
1080	1976.0	1977.8	1979.6	1981.4	1983.2	1985.0	1986.8	1988.6	1990.4	1992.2
1090	1994.0	1995.8	1997.6	1999.4	2001.2	2003.0	2004.8	2006.6	2008.4	2010.2
1100	2012.0	2013.8	2015.6	2017.4	2019.2	2021.0	2022.8	2024.6	2026.4	2028.2
1110	2030.0	2031.8	2033.6	2035.4	2037.2	2039.0	2040.8	2042.6	2044.4	2046.2
1120	2048.0	2049.8	2051.6	2053.4	2055.2	2057.0	2058.8	2060.6	2062.4	2064.2
1130	2066.0	2067.8	2069.6	2071.4	2073.2	2075.0	2076.8	2078.6	2080.4	2082.2
1140	2084.0	2085.8	2087.6	2089.4	2091.2	2093.0	2094.8	2096.6	2098.4	2100.2
1150	2102.0	2103.8	2105.6	2107.5	2109.2	2111.0	2112.8	2114.6	2116.4	2118.2
1160	2120.0	2121.8	2123.6	2125.4	2127.2	2129.0	2130.8	2132.6	2134.4	2136.2
1170	2138.0	2139.8	2141.6	2143.4	2145.2	2147.0	2148.8	2150.6	2152.4	2154.2
1180	2156.0	2157.8	2159.6	2161.4	2163.2	2165.0	2166.8	2168.6	2170.4	2172.2
1190	2174.0	2175.8	2177.6	2179.4	2181.2	2183.0	2184.8	2186.6	2188.4	2190.2
1200	2192.0	2193.8	2195.6	2197.4	2199.2	2201.0	2202.8	2204.6	2206.4	2208.2
1210	2210.0	2211.8	2213.6	2215.4	2217.2	2219.0	2220.8	2222.6	2224.4	2226.2
1220	2228.0	2229.8	2231.6	2233.4	2235.2	2237.0	2238.8	2240.6	2242.4	2244.2
1230	2246.0	2247.8	2249.6	2251.4	2253.2	2255.0	2256.8	2258.6	2260.4	2262.2
1240	2264.0	2265.8	2267.6	2269.4	2271.2	2273.0	2274.8	2276.6	2278.4	2280.2
1250	2282.0	2283.8	2285.6	2287.4	2289.2	2291.0	2292.8	2294.6	2296.4	2298.2
1260	2300.0	2301.8	2303.6	2305.4	2307.2	2309.0	2310.8	2312.6	2314.4	2316.2
1270	2318.0	2319.8	2321.6	2323.4	2325.2	2327.0	2328.8	2330.6	2332.4	2334.2
1280	2336.0	2337.8	2339.6	2341.4	2343.2	2345.0	2346.8	2348.6	2350.4	2352.2
1290	2354.0	2355.8	2357.6	2359.4	2361.2	2363.0	2364.8	2366.6	2368.4	2370.2
1300	2372.0	2373.8	2375.6	2377.4	2379.2	2381.0	2382.8	2384.6	2386.4	2388.2
1310	2390.0	2391.8	2393.6	2395.4	2397.2	2399.0	2400.8	2402.6	2404.4	2406.2
1320	2408.0	2409.8	2411.6	2413.4	2415.2	2417.0	2418.8	2420.6	2422.4	2424.2
1330	2426.0	2427.8	2429.6	2431.4	2433.2	2435.0	2436.8	2438.6	2440.4	2442.2
1340	2444.0	2445.8	2447.6	2449.4	2451.2	2453.0	2454.8	2456.6	2458.4	2460.2
1350	2462.0	2463.8	2465.6	2467.4	2469.2	2471.0	2472.8	2474.6	2476.4	2478.2
1360	2480.0	2481.8	2483.6	2485.4	2487.2	2489.0	2490.8	2492.6	2494.4	2496.2
1370	2498.0	2499.8	2501.6	2503.4	2505.2	2507.0	2508.8	2510.6	2512.4	2514.2
1380	2516.0	2517.8	2519.6	2521.4	2523.2	2525.0	2526.8	2528.6	2530.4	2532.2
1390	2534.0	2535.8	2537.6	2539.4	2541.2	2543.0	2544.8	2546.6	2548.4	2550.2
1400	2552.0	2553.8	2555.6	2557.4	2559.2	2561.0	2562.8	2564.6	2566.4	2568.2
1410	2570.0	2571.8	2573.6	2575.4	2577.2	2579.0	2580.8	2582.6	2584.4	2586.2
1420	2588.0	2589.8	2591.6	2593.4	2595.2	2597.0	2598.8	2600.6	2602.4	2604.2
1430	2606.0	2607.8	2609.6	2611.4	2613.2	2615.0	2616.8	2618.6	2620.4	2622.2
1440	2624.0	2625.8	2627.6	2629.4	2631.2	2633.0	2634.8	2636.6	2638.4	2640.2
1450	2642.0	2643.8	2645.6	2647.4	2649.2	2651.0	2652.8	2654.6	2656.4	2658.2
1460	2660.0	2661.8	2663.6	2665.4	2667.2	2669.0	2670.8	2672.6	2674.4	2676.2
1470	2678.0	2679.8	2681.6	2683.4	2685.2	2687.0	2688.8	2690.6	2692.4	2694.2
1480	2696.0	2697.8	2699.6	2701.4	2703.2	2705.0	2706.8	2708.6	2710.4	2712.2
1490	2714.0	2715.8	2717.6	2719.4	2721.2	2723.0	2724.8	2726.6	2728.4	2730.2

APPENDIX D (Continued)
Temperatures—Celsius to Fahrenheit
Conversion Table for temperatures above 0° C

Temp. °C	0	1	2	3	4	5	6	7	8	9
1500	2732.0	2733.8	2735.6	2737.4	2739.2	2741.0	2742.8	2744.6	2746.4	2748.2
1510	2750.0	2751.8	2753.6	2755.4	2757.2	2759.0	2760.8	2762.6	2764.4	2766.2
1520	2768.0	2769.8	2771.6	2773.4	2775.2	2777.0	2778.8	2780.6	2782.4	2784.2
1530	2786.0	2787.8	2789.6	2791.4	2793.2	2795.0	2796.8	2798.6	2800.4	2802.2
1540	2804.0	2805.8	2807.6	2809.4	2811.2	2813.0	2814.8	2816.6	2818.4	2820.2
1550	2822.0	2823.8	2825.6	2827.4	2829.2	2831.0	2832.8	2834.6	2836.4	2838.2
1560	2840.0	2841.8	2843.6	2845.4	2847.2	2849.0	2850.8	2852.6	2854.4	2856.2
1570	2858.0	2859.8	2861.6	2863.4	2865.2	2867.0	2868.8	2870.6	2872.4	2874.2
1580	2876.0	2877.8	2879.6	2881.4	2883.2	2885.0	2886.8	2888.6	2890.4	2892.2
1590	2894.0	2895.8	2897.6	2899.4	2901.2	2903.0	2904.8	2906.6	2908.4	2910.2
1600	2912.0	2913.8	2915.6	2917.4	2919.2	2921.0	2922.8	2924.6	2926.4	2928.2
1610	2930.0	2931.8	2933.6	2935.4	2937.2	2939.0	2940.8	2942.6	2944.4	2946.2
1620	2948.0	2949.8	2951.6	2953.4	2955.2	2957.0	2958.8	2960.6	2962.4	2964.2
1630	2966.0	2967.8	2969.6	2971.4	2973.2	2975.0	2976.8	2978.6	2980.4	2982.2
1640	2984.0	2985.8	2987.6	2989.4	2991.2	2993.0	2994.8	2996.6	2998.4	3000.2
1650	3002.0	3003.8	3005.6	3007.4	3009.2	3011.0	3012.8	3014.6	3016.4	3018.2
1660	3020.0	3021.8	3023.6	3025.4	3027.2	3029.0	3030.8	3032.6	3034.4	3036.2
1670	3038.0	3039.8	3041.6	3043.4	3045.2	3047.0	3048.8	3050.6	3052.4	3054.2
1680	3056.0	3057.8	3059.6	3061.4	3063.2	3065.0	3066.8	3068.6	3070.4	3072.2
1690	3074.0	3075.8	3077.6	3079.4	3081.2	3083.0	3084.8	3086.6	3088.4	3090.2
1700	3092.0	3093.8	3095.6	3097.4	3099.2	3101.0	3102.8	3104.6	3106.4	3108.2
1710	3110.0	3111.8	3113.6	3115.4	3117.2	3119.0	3120.8	3122.6	3124.4	3126.2
1720	3128.0	3129.8	3131.6	3133.4	3135.2	3137.0	3138.8	3140.6	3142.4	3144.2
1730	3146.0	3147.8	3149.6	3151.4	3153.2	3155.0	3156.8	3158.6	3160.4	3162.2
1740	3164.0	3165.8	3167.6	3169.4	3171.2	3173.0	3174.8	3176.6	3178.4	3180.2
1750	3182.0	3183.8	3185.6	3187.4	3189.2	3191.0	3192.8	3194.6	3196.4	3198.2
1760	3200.0	3201.8	3203.6	3205.4	3207.2	3209.0	3210.8	3212.6	3214.4	3216.2
1770	3218.0	3219.8	3221.6	3223.4	3225.2	3227.0	3228.8	3230.6	3232.4	3234.2
1780	3236.0	3237.8	3239.6	3241.4	3243.2	3245.0	3246.8	3248.6	3250.4	3252.2
1790	3254.0	3255.8	3257.6	3259.4	3261.2	3263.0	3264.8	3266.6	3268.4	3270.2
1800	3272.0	3273.8	3275.6	3277.4	3279.2	3281.0	3282.8	3284.6	3286.4	3288.2
1810	3290.0	3291.8	3293.6	3295.4	3297.2	3299.0	3300.8	3302.6	3304.4	3306.2
1820	3308.0	3309.8	3311.6	3313.4	3315.2	3317.0	3318.8	3320.6	3322.4	3324.2
1830	3326.0	3327.8	3329.6	3331.4	3332.2	3335.0	3336.8	3338.6	3340.4	3342.2
1840	3344.0	3345.8	3347.6	3349.4	3351.2	3353.0	3354.8	3356.6	3358.4	3360.2
1850	3362.0	3363.8	3365.6	3367.4	3369.2	3371.0	3372.8	3374.6	3376.4	3378.2
1860	3380.0	3381.8	3383.6	3385.4	3387.2	3389.0	3390.8	3392.6	3394.4	3396.2
1870	3398.0	3399.8	3401.6	3403.4	3405.2	3407.0	3408.8	3410.6	3412.4	3414.2
1880	3416.0	3417.8	3419.6	3421.4	3423.2	3425.0	3426.8	3428.6	3430.4	3432.2
1890	3434.0	3435.8	3437.6	3439.4	3441.2	3443.0	3444.8	3446.6	3448.4	3450.2
1900	3452.0	3453.8	3455.6	3457.4	3459.2	3461.0	3462.8	3464.6	3466.4	3468.2
1910	3470.0	3471.8	3473.6	3475.4	3477.2	3479.0	3480.8	3482.6	3484.4	3486.2
1920	3488.0	3489.8	3491.6	3493.4	3495.2	3497.0	3498.8	3500.6	3502.4	3504.2
1930	3506.0	3507.8	3509.6	3511.4	3513.2	3515.0	3516.8	3518.6	3520.4	3522.2
1940	3524.0	3525.8	3527.6	3529.4	3531.2	3533.0	3534.8	3536.6	3538.4	3540.2
1950	3542.0	3543.8	3545.6	3547.4	3549.2	3551.0	3552.8	3554.6	3556.4	3558.2
1960	3560.0	3561.8	3563.6	3565.4	3567.2	3569.0	3570.8	3572.6	3574.4	3576.2
1970	3578.0	3579.8	3581.6	3583.4	3585.2	3587.0	3588.8	3590.6	3592.4	3594.2
1980	3596.0	3597.8	3599.6	3601.4	3603.2	3605.0	3606.8	3608.6	3610.4	3612.2
1990	3614.0	3615.8	3617.6	3619.4	3621.2	3623.0	3624.8	3626.6	3628.4	3630.2

APPENDIX D (Continued)
Temperatures—Celsius to Fahrenheit
Conversion Table for Temperatures above 0° C

Temp. °C	0	1	2	3	4	5	6	7	8	9
2000	3632.0	3633.8	3635.6	3637.4	3639.2	3641.0	3642.8	3644.6	3646.4	3648.2
2010	3650.0	3651.8	3653.6	3655.4	3657.2	3659.0	3660.8	3662.6	3664.2	3666.2
2020	3668.0	3669.8	3671.6	3673.4	3675.2	3677.0	3678.8	3680.6	3682.4	3684.2
2030	3686.0	3687.8	3689.6	3691.4	3693.2	3695.0	3696.8	3698.6	3700.4	3702.2
2040	3704.0	3705.8	3707.6	3709.4	3711.2	3713.0	3714.8	3716.6	3718.4	3720.2
2050	3722.0	3723.8	3725.6	3727.4	3729.2	3731.0	3732.8	3734.6	3736.4	3738.2
2060	3740.0	3741.8	3743.8	3745.4	3747.2	3749.0	3750.8	3752.6	3754.4	3756.2
2070	3758.0	3759.8	3761.6	3763.4	3765.2	3767.0	3768.8	3770.6	3772.4	3774.2
2080	3776.0	3777.8	3779.6	3781.4	3783.2	3785.0	3786.8	3788.6	3790.4	3792.2
2090	3794.0	3795.8	3797.6	3799.4	3801.2	3803.0	3804.8	3806.6	3808.4	3810.2
2100	3812.0	3813.8	3815.6	3817.4	3819.2	3821.0	3822.8	3824.6	3826.4	3828.2
2110	3830.0	3831.8	3833.6	3835.4	3837.2	3839.0	3840.8	3842.6	3844.4	3846.2
2120	3848.0	3849.8	3851.6	3853.4	3855.2	3857.0	3858.8	3860.6	3862.4	3664.2
2130	3866.0	3867.8	3869.6	3871.4	3873.2	3875.0	3876.8	3878.6	3880.4	3882.2
2140	3884.0	3885.8	3887.6	3889.4	3891.2	3893.0	3894.8	3896.6	3898.4	3900.2
2150	3902.0	3903.8	3905.6	3907.4	3909.2	3911.0	3912.8	3914.6	3916.4	3918.2
2160	3920.0	3921.8	3923.6	3925.4	3927.2	3929.0	3930.8	3932.6	3934.4	3936.2
2170	3938.0	3939.8	3941.6	3943.4	3945.2	3947.0	3948.8	3950.6	3952.4	3954.2
2180	3956.0	3957.8	3959.6	3961.4	3963.2	3965.0	3966.8	3968.6	3970.4	3972.2
2190	3974.0	3975.8	3977.6	3979.4	3981.2	3983.0	3984.8	3986.6	3988.4	3990.2
2200	3992.0	3993.8	3995.6	3997.4	3999.2	4001.0	4002.8	4004.6	4006.4	4008.2
2210	4010.0	4011.8	4013.6	4015.4	4017.2	4019.0	4020.8	4022.6	4024.4	4026.2
2220	4028.0	4029.8	4031.6	4033.4	4035.2	4037.0	4038.8	4040.6	4042.4	4044.2
2230	4046.0	4047.8	4049.6	4051.4	4053.2	4055.0	4056.8	4058.6	4060.4	4062.2
2240	4064.0	4065.8	4067.6	4069.4	4071.2	4073.0	4074.8	4076.6	4078.4	4080.2
2250	4082.0	4083.8	4085.6	4087.4	4089.2	4091.0	4092.8	4094.6	4096.4	4098.2
2260	4100.0	4101.8	4103.6	4105.4	4107.2	4109.0	4110.8	4112.6	4114.4	4116.2
2270	4118.0	4119.8	4121.6	4123.4	4125.2	4127.0	4128.8	4130.6	4132.4	4134.2
2280	4136.0	4137.8	4139.6	4141.4	4143.2	4145.0	4146.8	4148.6	4150.4	4152.2
2290	4154.0	4155.8	4157.6	4159.4	4161.2	4163.0	4164.8	4166.6	4168.4	4170.2
2300	4172.0	4173.8	4175.6	4177.4	4179.2	4181.0	4182.8	4184.6	4186.4	4188.2
2310	4190.0	4191.8	4193.6	4195.4	4197.2	4199.0	4200.8	4202.6	4204.4	4206.2
2320	4208.0	4209.8	4211.6	4213.4	4215.2	4217.0	4218.8	4220.6	4222.4	4224.2
2330	4226.0	4227.8	4229.6	4231.4	4233.2	4235.0	4236.8	4238.6	4240.4	4242.2
2340	4244.0	4245.8	4247.6	4249.4	4251.2	4253.0	4254.8	4256.6	4258.4	4260.2
2350	4262.0	4263.8	4265.6	4267.4	4269.2	4271.0	4272.8	4274.6	4276.4	4278.2
2360	4280.0	4281.8	4283.6	4285.4	4287.2	4289.0	4290.8	4292.6	4294.4	4296.2
2370	4298.0	4299.8	4301.6	4303.4	4305.2	4307.0	4308.8	4310.6	4312.4	4314.2
2380	4316.0	4317.8	4319.6	4321.4	4323.2	4325.0	4326.8	4328.6	4330.4	4332.2
2390	4334.0	4335.8	4337.6	4339.4	4341.2	4343.0	4344.8	4346.6	4348.4	4350.2

APPENDIX D (Continued)
Metric Bibliography

American Society for Testing and Materials. *Metric Practice Guide.* Bulletin No. E 380-72. Philadelphia: American Society for Testing and Materials, 1973, 34 pp.

Donovan, Frank R. *Prepare Now for a Metric Future.* New York, New York: Weybright & Talley, 1971, 212 pp.

Jones, M. J. B. *A Guide to Metrication.* Oxford, England: Pergamon Press Limited, 1969, 156 pp.

U.S. Metric Study Interim Reports. *A Metric America.* Bulletin No. C13.10:345. Washington, D.C.: Superintendent of Documents, U.S. Government Printing Office, 1970, 192 pp. (The first of 13 reports by the U.S. Metric Study Committee. The others are available from the same source.)

APPENDIX E
Selected Physical Properties of Some Common Glasses

Type	Youngs Modulus (10^6 PSI)	Modulus of Rigidity (10^6 PSI)	Specific Gravity	Refraction Index	Knoop Hardness (50 gram load)	Working Annealing	Temperature (°F) Softening
Soda-lime silica (Plate)	10.0	4.3	2.47	1.51	570	1020	1350
Borosilicate	9.2	3.8	2.23	1.47	—	1050	1500
Lead-Alkali Silicate	9.0	3.7	2.85	1.54	450	815	1160
Fused silica	10.0	4.5	2.20	1.46	650	2100	3000
96% Silica	9.6	4.1	2.18	1.46	590	1670	2700
Alumina Silicate	12.7	5.2	2.53	1.53	640	1320	1675

APPENDIX F

Selected Strength Properties of Some Hardwoods and Softwoods[1]

(Results of Tests on Green and Air-dry Specimens)[2]

Common Name of Species	Moisture Content, Pct.	Specific Gravity	Static Bending — Modulus of Rupture, PSI	Static Bending — Modulus of Elasticity, 1,000 PSI	Static Bending — Work to Maximum Load, in/lb PCI	Compression Parallel to Grain-Maximum Crush Strength, PSI	Shear Parallel to Grain-Maximum Shear Strength, PSI	Tension Perpendicular to Grain-Maximum Tensile Strength, PSI	Side Hardness, lb.	Machineability, Surface Quality[3]	Typical Uses of Species
HARDWOODS											
Afrormosia	62 / 12	.66 / .74	14,820 / 18,430	1,776 / 1,937	19.5 / 18.5	7,488 / 9,936	1,672 / 2,086	... / ...	1,600 / 1,560	A	Boat construction, veneers; teak substitute
Ash, White	42 / 12	.55 / .60	9,600 / 15,400	1,460 / 1,770	16.6 / 17.6	3,990 / 7,410	1,380 / 1,950	590 / 940	960 / 1,320	G	Tool handles, oars, vehicle parts, sporting equipment.
Aspen, Quaking	94 / 12	.35 / .38	5,100 / 8,400	860 / 1,180	6.4 / 7.6	2,140 / 4,250	660 / 850	230 / 260	300 / 350	G	Crating, matches, veneer, and pulpwood.
Basswood, American	105 / 12	.32 / .37	5,000 / 8,700	1,040 / 1,460	5.3 / 7.2	2,220 / 4,730	600 / 990	280 / 350	250 / 410	A	Woodenware, venetian blinds, and boxes
Birch, Yellow	67 / 12	.55 / .62	8,300 / 16,600	1,500 / 2,010	16.1 / 20.8	3,380 / 8,170	1,110 / 1,880	430 / 920	780 / 1,260	A	Furniture, woodenware; plywood doors, cabinets, toys, and cooperage
Butternut	104 / 12	.36 / .38	5,400 / 8,100	970 / 1,180	8.2 / 8.2	2,420 / 5,110	760 / 1,170	430 / 440	390 / 490	A	Woodenware, furniture, toys, and instrument cases.
Cherry, Black	55 / 12	.47 / .50	8,000 / 12,300	1,310 / 1,490	12.8 / 11.4	3,540 / 7,110	1,130 / 1,700	570 / 560	660 / 950	A	Furniture, caskets, woodenware and gunstocks
Chestnut, American	122 / 12	.40 / .43	5,600 / 8,600	930 / 1,230	7.0 / 6.5	2,740 / 5,320	800 / 1,080	440 / 460	420 / 540	G	Railway ties, furniture, caskets and core stock for veneer panels
Ebony, African	GR / 12	.90 / 1.08	... / 26,030	... / 2,739	... / 28.1	... / 12,816	... / 2,473	... / ...	... / 3,220	A	Musical instrument parts, carvings, and turnings.

APPENDIX F (Continued)

Common Name of Species	Moisture Content, Pct.	Specific Gravity	Static Bending			Compression Parallel to Grain-Maximum Crush Strength, PSI	Shear Parallel to Grain-Maximum Shear Strength, PSI	Tension Perpendicular to Grain-Maximum Tensile Strength, PSI	Side Hardness, lb.	Machineability, Surface Quality[3]	Typical Uses of Species
			Modulus of Rupture, PSI	Modulus of Elasticity, 1,000 PSI	Work to Maximum Load, in/lb PCI						
HARDWOODS (Cont'd.)											
Elm, American	89 12	.46 .50	7,200 11,800	1,110 1,340	11.8 13.0	2,910 5,520	1,000 1,510	590 660	620 830	P	Crating, furniture, farm implements, and veneer
Hickory, Shagbark	60 12	.64 .72	11,000 20,200	1,570 2,160	23.7 25.8	4,580 9,210	1,520 2,430	... ...	... ...	G	Tool handles, athletic equipment, dowels, ladder rings, and furniture
Mahogany, American	67 12	.45 .50	9,280 11,640	1,280 1,510	9.6 7.9	4,510 6,630	1,310 1,290	... ...	700 810	G	Interior trim, panelling, furniture, boat construction
Mahogany, Luan	GR 12	.44 .50	7,700 11,300	1,380 1,630	... ...	3,700 5,890	930 1,220	... ...	570 680	G	Interior trim, panelling, furniture, boat construction
Maple, Sugar	58 12	.56 .63	9,400 15,800	1,550 1,830	13.3 16.5	4,020 7,830	1,460 2,330	... ...	970 1,450	A	Flooring, furniture, handles, and woodenware
Oak, Northern Red	80 12	.56 .63	8,300 14,300	1,350 1,820	13.2 14.5	3,440 6,760	1,210 1,780	750 800	1,000 1,290	G	Flooring, railroad ties, furniture, woodenware, and handles.
Oak, White	68 12	.60 .68	8,300 15,200	1,250 1,780	11.6 14.8	3,560 7,440	1,250 2,000	770 800	1,060 1,360	G	White oak desirable for cooperage
Rosewood, Brazillian	GR 12	.80 .88	14,140 18,970	1,840 1,880	13.2 ...	5,510 9,600	2,360 2,110	... ...	2,440 2,720	A	Veneers, fine furniture, quality handles
Sweetgum	115 12	.46 .52	7,100 12,500	1,200 1,640	10.1 11.9	3,040 6,320	990 1,600	540 760	600 850	A	Crating, furniture, cabinets, and millwork
Sycamore	83 12	.46 .49	6,500 10,000	1,060 1,420	7.5 8.5	2,920 5,380	1,000 1,470	630 720	610 770	P	Cooperage, furniture, food boxes, and handles
Teak, Asian	67 12	.57 .65	10,980 10,780	1,510 1,590	10.8 10.1	5,470 7,110	1,290 1,480	... ...	1,070 1,030	G	Veneers, fine furniture, boat construction

APPENDIX F (Continued)
Selected Strength Properties of Some Hardwoods and Softwoods[1]
(Results of Tests on Green and Air-dry Specimens)[2]

Common Name of Species	Moisture Content, Pct.	Specific Gravity	Static Bending — Modulus of Rupture, PSI	Static Bending — Modulus of Elasticity, 1,000 PSI	Static Bending — Work to Maximum Load, in/lb PCI	Compression Parallel to Grain-Maximum Crush Strength, PSI	Shear Parallel to Grain-Maximum Shear Strength, PSI	Tension Perpendicular to Grain-Maximum Tensile Strength, PSI	Side Hardness, lb.	Machineability, Surface Quality[3]	Typical Uses of Species
HARDWOODS (Cont'd.)											
Walnut, Black	81 / 12	.51 / .55	9,500 / 14,600	1,420 / 1,680	14.6 / 10.7	4,300 / 7,580	1,220 / 1,370	570 / 690	900 / 1,010	G	Furniture, gunstocks, veneers, and cabinets
Yellow Poplar	83 / 12	.40 / .42	6,000 / 10,100	1,220 / 1,580	7.5 / 8.8	2,620 / 5,540	790 / 1,190	510 / 540	440 / 540	A	Furniture, cooperage, crating, and plywood core stock
SOFTWOODS											
Cedar, Northern White	55 / 12	.29 / .31	4,200 / 6,500	640 / 800	5.7 / 4.8	1,990 / 3,960	620 / 850	240 / 240	230 / 320	G	Boats, posts, poles, and woodenware
Cedar, Western Red	37 / 12	.31 / .33	5,100 / 7,700	920 / 1,120	5.0 / 5.8	2,750 / 5,020	710 / 860	230 / 220	270 / 350	G	Millwork, exterior siding, boats, pencils, and crating
Douglas Fir, Coast Type	38 / 12	.45 / .48	7,600 / 12,200	1,570 / 1,950	7.6 / 9.8	3,860 / 7,430	930 / 1,160	300 / 340	500 / 710	A	Lumber, millwork, sash, and doors
Fir, White	115 / 12	.35 / .37	5,700 / 9,300	1,030 / 1,380	5.1 / 6.7	2,710 / 5,350	750 / 930	290 / 260	330 / 440	A	Lumber, millwork, crating, doors, and exterior siding
Hemlock, Eastern	111 / 12	.38 / .40	6,400 / 8,900	1,070 / 1,200	6.7 / 6.8	3,080 / 5,410	850 / 1,060	230 / ...	400 / 500	A / P	Lumber, pulpwood, and crating
Pine, Eastern White	73 / 12	.34 / .35	4,900 / 8,600	990 / 1,240	5.2 / 6.8	2,440 / 4,800	680 / 900	250 / 310	290 / 380	G	Lumber, millwork, foundry patterns, crating, and matches. Jack pine mainly for pulpwood, fuel and box lumber; red pine for cabin logs and piling.
Pine, Jack	60 / 12	.40 / .48	6,000 / 9,900	1,070 / 1,350	7.2 / 8.3	2,950 / 5,660	750 / 1,170	360 / 420	400 / 570	A	
Pine, Loblolly	81 / 12	.47 / .51	7,300 / 12,800	1,410 / 1,800	8.2 / 10.4	3,490 / 7,080	850 / 1,370	260 / 470	450 / 690	A	

APPENDIX F (Continued)

Common Name of Species	Moisture Content, Pct.	Specific Gravity	Static Bending — Modulus of Rupture, PSI	Static Bending — Modulus of Elasticity, 1,000 PSI	Static Bending — Work to Maximum Load, in/lb PCI	Compression Parallel to Grain-Maximum Crush Strength, PSI	Shear Parallel to Grain-Maximum Shear Strength, PSI	Tension Perpendicular to Grain-Maximum Tensile Strength, PSI	Side Hardness, lb.	Machineability, Surface Quality[3]	Typical Uses of Species
SOFTWOODS (Cont'd.)											
Pine, Ponderosa	91 12	.38 .40	5,000 9,200	970 1,260	5.1 6.6	2,400 5,270	680 1,160	290 400	310 450	A	Lumber, millwork, foundry patterns, crating, and matches.
Pine, Red	92 12	.41 .44	5,800 11,000	1,280 1,630	6.1 9.9	2,730 6,070	690 1,210	300 460	340 560	A	Jack pine mainly for pulpwood, fuel and box lumber; red pine for cabin logs and piling.
Pine, Western White	54 12	.36 .38	5,200 9,500	1,170 1,510	5.0 8.8	2,650 5,620	640 850	260 ...	310 370	G	
Redwood	112 12	.38 .40	7,500 10,000	1,180 1,340	7.4 6.9	4,200 6,150	800 940	260 240	410 480	A	Lumber, outdoor furniture, silos, and fencing
Spruce, Black	38 12	.38 .40	5,400 10,300	1,060 1,530	7.4 10.5	2,570 5,320	660 1,030	100 ...	370 520	A	Pulpwood, millwork, and crating
Spruce, White	50 12	.37 .40	5,600 9,800	1,070 1,340	6.0 7.7	2,570 5,470	690 1,080	220 360	320 480	A	
Spruce, Englemann	80 12	.32 .34	4,500 8,700	960 1,280	5.1 6.4	2,190 4,770	590 1,030	240 350	260 350	A	Lumber, mining timber, and poles
Tamarack	52 12	.49 .53	7,200 11,600	1,240 1,640	7.2 7.1	3,480 7,160	860 1,280	260 400	380 590	P	Pulpwood, fuel, poles, and crating

[1] Based upon data from *Wood Handbook*, and *Properties of Imported Tropical Woods*, both listed in the bibliography in Chapter 6.
[2] The values in the first line for each species indicate green wood, based upon volume; those in the second line are for seasoned wood adjusted to an average air-dry moisture content of 12%, based upon weight.
[3] Machinability as a measure of surface quality: G = good or smooth surface; A = average surface; P = poor surface, requiring special care.

APPENDIX G
Percentage Reduction of Area for Tensile Test Specimens
(Originally .505″, or .506″, or .504″ Diameter)

Reduced Dia. In.	Area	% R. A. 0.505 In.	% R. A. 0.506 In.	% R. A. 0.504 In.	Reduced Dia. In.	Area	% R. A. 0.505 In.	% R. A. 0.506 In.	% R. A. 0.504 In.
0.251	0.0494	75.3	75.4	75.2	0.315	0.0779	61.1	61.2	61.0
0.252	0.0498	75.1	75.2	75.0	0.316	0.0784	60.8	61.0	60.7
0.253	0.0502	74.9	75.0	74.8	0.317	0.0789	60.6	60.7	60.5
0.254	0.0506	74.7	74.8	74.6	0.318	0.0794	60.3	60.5	60.2
0.255	0.0510	74.5	74.6	74.4	0.319	0.0799	60.1	60.2	60.0
0.256	0.0514	74.3	74.4	74.2					
0.257	0.0518	74.1	74.2	74.0	0.320	0.0804	59.8	60.0	59.7
0.258	0.0522	73.9	74.0	73.8	0.321	0.0809	59.6	59.8	59.4
0.259	0.0526	73.7	73.8	73.6	0.322	0.0814	59.4	59.5	59.2
					0.323	0.0819	59.1	59.3	58.9
0.260	0.0530	73.5	73.6	73.4	0.324	0.0824	58.8	59.0	58.7
0.261	0.0535	73.3	73.4	73.2	0.325	0.0829	58.6	58.8	58.4
0.262	0.0539	73.0	73.2	73.0	0.326	0.0834	58.3	58.5	58.2
0.263	0.0543	72.9	73.0	72.8	0.327	0.0839	58.1	58.3	57.9
0.264	0.0547	72.7	72.8	72.6	0.328	0.0844	57.8	58.0	57.7
0.265	0.0551	72.5	72.6	72.4	0.329	0.0850	57.5	57.7	57.4
0.266	0.0555	72.3	72.4	72.2					
0.267	0.0559	71.9	72.2	72.0	0.330	0.0855	57.3	57.5	57.1
0.268	0.0564	71.8	71.9	71.7	0.331	0.0860	57.0	57.2	56.9
0.269	0.0568	71.6	71.7	71.5	0.332	0.0865	56.8	57.0	56.6
					0.333	0.0870	56.5	56.7	56.4
0.270	0.0572	71.4	71.5	71.3	0.334	0.0876	56.2	56.4	56.1
0.271	0.0576	71.2	71.3	71.1	0.335	0.0881	56.0	56.2	55.8
0.272	0.0581	71.0	71.1	70.9	0.336	0.0886	55.7	55.9	55.6
0.273	0.0585	70.8	70.9	70.7	0.337	0.0891	55.5	55.7	55.3
0.274	0.0589	70.6	70.7	70.5	0.338	0.0897	55.2	55.4	55.0
0.275	0.0593	70.4	70.5	70.3	0.339	0.0902	54.9	55.1	54.8
0.276	0.0598	70.1	70.2	70.0					
0.277	0.0602	69.9	70.0	69.8	0.340	0.0907	54.7	54.9	54.5
0.278	0.0606	69.7	69.9	69.6	0.341	0.0913	54.4	54.6	54.2
0.279	0.0611	69.5	69.6	69.4	0.342	0.0918	54.1	54.3	54.0
					0.343	0.0924	53.8	54.0	53.7
0.280	0.0615	69.3	69.4	69.2	0.344	0.0929	53.6	53.8	53.4
0.281	0.0620	69.0	69.2	69.0	0.345	0.0934	53.3	53.5	53.2
0.282	0.0624	68.8	69.0	68.7	0.346	0.0940	53.0	53.2	52.9
0.283	0.0629	68.6	68.7	68.5	0.347	0.0945	52.8	53.0	52.6
0.284	0.0633	68.4	68.5	68.3	0.348	0.0951	52.5	52.7	52.3
0.285	0.0637	68.2	68.3	68.1	0.349	0.0956	52.2	52.4	52.1
0.286	0.0642	67.9	68.1	67.8					
0.287	0.0646	67.7	67.9	67.6	0.350	0.0962	51.9	52.1	51.8
0.288	0.0651	67.5	67.6	67.4	0.351	0.0967	51.7	51.9	51.5
0.289	0.0655	67.3	67.4	67.2	0.352	0.0973	51.4	51.6	51.2
					0.353	0.0978	51.1	51.3	51.0
0.290	0.0660	67.0	67.2	66.9	0.354	0.0984	50.8	51.0	50.7
0.291	0.0665	66.8	66.9	66.7	0.355	0.0989	50.6	50.8	50.4
0.292	0.0670	66.5	66.7	66.4	0.356	0.0995	50.3	50.5	50.1
0.293	0.0674	66.3	66.5	66.2	0.357	0.1000	50.0	50.2	49.9
0.294	0.0679	66.1	66.2	66.0	0.358	0.1006	49.8	50.0	49.6
0.295	0.0683	65.9	66.0	65.8	0.359	0.1012	49.5	49.7	49.3
0.296	0.0688	65.6	65.8	65.5					
0.297	0.0692	65.4	65.6	65.3	0.360	0.1017	49.2	49.4	49.0
0.298	0.0697	65.2	65.3	65.1	0.361	0.1023	48.9	49.1	48.7
0.299	0.0702	64.9	65.1	64.8	0.362	0.1029	48.6	48.8	48.4
					0.363	0.1034	48.4	48.6	48.2
0.300	0.0707	64.7	64.8	64.6	0.364	0.1040	48.1	48.3	47.9
0.301	0.0712	64.4	64.6	64.3	0.365	0.1046	47.8	48.0	47.6
0.302	0.0716	64.2	64.4	64.1	0.366	0.1052	47.5	47.7	47.3
0.303	0.0721	64.0	64.1	63.9	0.367	0.1057	47.2	47.4	47.0
0.304	0.0725	63.8	63.9	63.7	0.368	0.1063	46.9	47.1	46.7
0.305	0.0730	63.5	63.7	63.4	0.369	0.1069	46.6	46.8	46.4
0.306	0.0735	63.3	63.4	63.2					
0.307	0.0740	63.0	63.2	62.9	0.370	0.1075	46.3	46.5	46.1
0.308	0.0745	62.8	62.9	62.7	0.371	0.1081	46.0	46.2	45.8
0.309	0.0749	62.6	62.7	62.5	0.372	0.1086	45.8	46.0	45.6
					0.373	0.1092	45.5	45.7	45.3
0.310	0.0754	62.3	62.5	62.2	0.374	0.1098	45.2	45.4	45.0
0.311	0.0759	62.1	62.2	62.0	0.375	0.1104	44.9	45.1	44.5
0.312	0.0764	61.8	62.0	61.7	0.376	0.1110	44.6	44.8	44.4
0.313	0.0769	61.6	61.7	61.5	0.377	0.1116	44.3	44.5	44.1
0.314	0.0774	61.3	61.5	61.2	0.378	0.1122	44.0	44.2	43.8

APPENDIX G (Continued)
Percentage Reduction of Area for Tensile Test Specimens
(Originally .505″, or .506″, or .504″ Diameter)

Reduced Dia. In.	Area	% R. A. 0.505 In.	% R. A. 0.506 In.	% R. A. 0.504 In.	Reduced Dia. In.	Area	% R. A. 0.505 In.	% R. A. 0.506 In.	% R. A. 0.504 In.
0.379	0.1128	43.7	43.9	43.5	0.443	0.1541	23.0	23.3	22.8
					0.444	0.1548	22.7	23.0	22.5
0.380	0.1134	43.4	43.6	43.2	0.445	0.1555	22.3	22.6	22.1
0.381	0.1140	43.1	43.3	42.9	0.446	0.1562	22.0	22.3	21.7
0.382	0.1146	42.8	43.0	42.6	0.447	0.1569	21.6	21.9	21.4
0.383	0.1152	42.5	42.7	42.3	0.448	0.1576	21.3	21.6	21.0
0.384	0.1158	42.2	42.4	42.0	0.449	0.1583	20.9	21.2	20.7
0.385	0.1164	41.9	42.1	41.7					
0.386	0.1170	41.6	41.8	41.4	0.450	0.1590	20.6	20.9	20.3
0.387	0.1176	41.3	41.5	41.1	0.451	0.1597	20.2	20.5	20.0
0.388	0.1182	41.0	41.2	40.8	0.452	0.1604	19.9	20.2	19.6
0.389	0.1188	40.7	40.9	40.5	0.453	0.1611	19.5	19.9	19.2
					0.454	0.1618	19.2	19.5	18.9
0.390	0.1194	40.4	40.6	40.2	0.455	0.1625	18.8	19.2	18.5
0.391	0.1200	40.1	40.3	39.9	0.456	0.1633	18.4	18.8	18.1
0.392	0.1206	39.8	40.0	39.6	0.457	0.1640	18.1	18.4	17.8
0.393	0.1213	39.4	39.7	39.2	0.458	0.1647	17.7	18.1	17.4
0.394	0.1219	39.1	39.4	38.9	0.459	0.1654	17.4	17.7	17.1
0.395	0.1225	38.8	39.1	38.6					
0.396	0.1231	38.5	38.8	38.3	0.460	0.1661	17.0	17.4	16.7
0.397	0.1237	38.2	38.5	38.0	0.461	0.1669	16.6	17.0	16.3
0.398	0.1244	37.9	38.1	37.6	0.462	0.1676	16.3	16.6	16.0
0.399	0.1250	37.6	37.8	37.3	0.463	0.1683	15.9	16.3	15.6
					0.464	0.1690	15.6	15.9	15.3
0.400	0.1256	37.3	37.5	37.0	0.465	0.1698	15.2	15.5	14.9
0.401	0.1262	37.0	37.2	36.7	0.466	0.1705	14.8	15.2	14.5
0.402	0.1269	36.6	36.9	36.4	0.467	0.1712	14.5	14.8	14.2
0.403	0.1275	36.3	36.6	36.1	0.468	0.1720	14.1	14.4	13.8
0.404	0.1281	36.0	36.3	35.8	0.469	0.1727	13.7	14.1	13.4
0.405	0.1288	35.7	35.9	35.4					
0.406	0.1294	35.4	35.6	35.1	0.470	0.1734	13.4	13.7	13.1
0.407	0.1301	35.0	35.3	34.8	0.471	0.1742	13.0	13.3	12.7
0.408	0.1307	34.7	35.0	34.5	0.472	0.1749	12.6	13.0	12.3
0.409	0.1313	34.4	34.7	34.2	0.473	0.1757	12.2	12.6	11.9
					0.474	0.1764	11.9	12.2	11.6
0.410	0.1320	34.1	34.3	33.8	0.475	0.1772	11.5	11.8	11.2
0.411	0.1326	33.8	34.0	33.5	0.476	0.1779	11.1	11.5	10.8
0.412	0.1333	33.4	33.7	33.2	0.477	0.1787	10.7	11.1	10.4
0.413	0.1339	33.1	33.4	32.9	0.478	0.1794	10.4	10.8	10.1
0.414	0.1346	32.8	33.0	32.5	0.479	0.1802	10.0	10.3	9.7
0.415	0.1352	32.5	32.8	32.2					
0.416	0.1359	32.1	32.4	31.9	0.480	0.1809	9.6	10.0	9.3
0.417	0.1365	31.8	32.1	31.6	0.481	0.1817	9.2	9.6	8.9
0.418	0.1372	31.5	31.7	31.2	0.482	0.1824	8.9	9.3	8.6
0.419	0.1378	31.2	31.4	30.9	0.483	0.1832	8.5	8.9	8.2
					0.484	0.1839	8.1	8.5	7.8
0.420	0.1385	30.8	31.1	30.6	0.485	0.1847	7.7	8.1	7.4
0.421	0.1392	30.5	30.7	30.3	0.486	0.1855	7.3	7.7	7.0
0.422	0.1398	30.2	30.4	29.9	0.487	0.1862	7.0	7.4	6.7
0.423	0.1405	29.8	30.1	29.6	0.488	0.1870	6.6	7.0	6.3
0.424	0.1411	29.5	29.8	29.3	0.489	0.1878	6.2	6.6	5.9
0.425	0.1418	29.2	29.5	28.9					
0.426	0.1425	28.8	29.1	28.6	0.490	0.1885	5.8	6.2	5.5
0.427	0.1432	28.5	28.8	28.2	0.491	0.1893	5.4	5.8	5.1
0.428	0.1438	28.2	28.5	27.9	0.492	0.1901	5.0	5.4	4.7
0.429	0.1445	27.8	28.1	27.6	0.493	0.1908	4.7	5.1	4.4
					0.494	0.1916	4.3	4.7	4.0
0.430	0.1452	27.5	27.8	27.4	0.495	0.1924	3.9	4.3	3.6
0.431	0.1458	27.2	27.5	26.9	0.496	0.1932	3.5	3.9	3.2
0.432	0.1465	26.8	27.1	26.6	0.497	0.1940	3.1	3.5	2.8
0.433	0.1472	26.5	26.8	26.2	0.498	0.1947	2.7	3.1	2.4
0.434	0.1479	26.1	26.4	25.9	0.499	0.1955	2.3	2.7	2.0
0.435	0.1486	25.8	26.1	25.5					
0.436	0.1493	25.4	25.7	25.2	0.500	0.1963	1.9	2.3	1.6
0.437	0.1499	25.1	25.4	24.9	0.501	0.1971	1.4	1.9	1.2
0.438	0.1506	24.8	25.1	24.5	0.502	0.1979	1.1	1.5	0.8
0.439	0.1513	24.4	24.7	24.2	0.503	0.1987	0.8	1.1	0.4
0.440	0.1520	24.1	24.4	23.9	0.504	0.1995	0.4	0.7	0.0
0.441	0.1527	23.7	24.0	23.5	0.505	0.2002	0.0	0.4	
0.442	0.1534	23.4	23.7	23.2	0.506	0.2010		0.0	

APPENDIX H

Diameter of Circles and Their Areas in Square Inches

Dia.	Area	Dia.	Area	Dia.	Area	Dia.	Area	Dia.	Area	Dia.	Area	Dia.	Area	Dia.	Area	Dia.	Area
.001	.0000008	.101	.008012	.201	.0317309	.301	.0711580	.401	.126293	.501	.197136	.601	.283687	.701	.385946	.801	.503913	.901	.637588
.002	.0000031	.102	.008171	.202	.0320474	.302	.0716316	.402	.126923	.502	.197923	.602	.284632	.702	.387048	.802	.505172	.902	.639004
.003	.0000071	.103	.008332	.203	.0323655	.303	.0721067	.403	.127556	.503	.198713	.603	.285578	.703	.388151	.803	.506432	.903	.640422
.004	.0000126	.104	.008495	.204	.0326852	.304	.0725835	.404	.128189	.504	.199504	.604	.286526	.704	.389256	.804	.507695	.904	.641841
.005	.0000196	.105	.008659	.205	.0330064	.305	.0730618	.405	.128825	.505	.200296	.605	.287476	.705	.390363	.805	.508958	.905	.643262
.006	.0000283	.106	.008825	.206	.0333292	.306	.0735417	.406	.129462	.506	.201090	.606	.288426	.706	.391471	.806	.510224	.906	.644684
.007	.0000385	.107	.008992	.207	.0336536	.307	.0740231	.407	.130100	.507	.201886	.607	.289379	.707	.392581	.807	.511490	.907	.646108
.008	.0000503	.108	.009161	.208	.0339795	.308	.0745061	.408	.130740	.508	.202683	.608	.290324	.708	.393692	.808	.512759	.908	.647534
.009	.0000636	.109	.009331	.209	.0343070	.309	.0749907	.409	.131382	.509	.203482	.609	.291289	.709	.394805	.809	.514029	.909	.648961
.010	.0000785	.110	.009508	.210	.0346361	.310	.0754769	.410	.132025	.510	.204282	.610	.292247	.710	.395920	.810	.515300	.910	.650389
.011	.0000950	.111	.009677	.211	.0349667	.311	.0759646	.411	.132670	.511	.205084	.611	.293206	.711	.397036	.811	.516574	.911	.651819
.012	.0001131	.112	.009852	.212	.0352990	.312	.0764539	.412	.133316	.512	.205887	.612	.294166	.712	.398153	.812	.517848	.912	.653251
.013	.0001327	.113	.0100287	.213	.0356328	.313	.0769448	.413	.133964	.513	.206692	.613	.295128	.713	.399273	.813	.519125	.913	.654685
.014	.0001539	.114	.0102070	.214	.0359681	.314	.0774372	.414	.134614	.514	.207499	.614	.296092	.714	.400393	.814	.520402	.914	.656120
.015	.0001767	.115	.0103869	.215	.0363051	.315	.0779313	.415	.135265	.515	.208307	.615	.297057	.715	.401516	.815	.521682	.915	.657556
.016	.0002011	.116	.0105683	.216	.0366436	.316	.0784268	.416	.135918	.516	.209117	.616	.298024	.716	.402640	.816	.522963	.916	.658994
.017	.0002270	.117	.0107513	.217	.0369837	.317	.0789240	.417	.136572	.517	.209928	.617	.298993	.717	.403765	.817	.524245	.917	.660434
.018	.0002545	.118	.0109359	.218	.0373253	.318	.0794227	.418	.137228	.518	.210741	.618	.299963	.718	.404892	.818	.525529	.918	.661875
.019	.0002835	.119	.0111220	.219	.0376685	.319	.0799230	.419	.137885	.519	.211556	.619	.300934	.719	.406021	.819	.526815	.919	.663318
.020	.0003142	.120	.0113097	.220	.0380133	.320	.0804249	.420	.138544	.520	.212372	.620	.301907	.720	.407151	.820	.528102	.920	.664762
.021	.0003464	.121	.0114990	.221	.0383597	.321	.0809284	.421	.139205	.521	.213189	.621	.302882	.721	.408283	.821	.529391	.921	.666208
.022	.0003801	.122	.0116898	.222	.0387076	.322	.0814334	.422	.139867	.522	.214008	.622	.303858	.722	.409416	.822	.530682	.922	.667655
.023	.0004155	.123	.0118823	.223	.0390571	.323	.0819399	.423	.140530	.523	.214829	.623	.304836	.723	.410551	.823	.531974	.923	.669105
.024	.0004524	.124	.0120763	.224	.0394082	.324	.0824481	.424	.141196	.524	.215651	.624	.305815	.724	.411687	.824	.533267	.924	.670555
.025	.0004909	.125	.0122718	.225	.0397608	.325	.0829578	.425	.141862	.525	.216475	.625	.306796	.725	.412825	.825	.534562	.925	.672007
.026	.0005309	.126	.0124690	.226	.0401150	.326	.0834691	.426	.142531	.526	.217301	.626	.307779	.726	.413965	.826	.535859	.926	.673461
.027	.0005726	.127	.0126677	.227	.0404708	.327	.0839820	.427	.143201	.527	.218128	.627	.308783	.727	.415106	.827	.537157	.927	.674916
.028	.0006158	.128	.0128679	.228	.0408282	.328	.0844964	.428	.143872	.528	.218956	.628	.309749	.728	.416249	.828	.538457	.928	.676373
.029	.0006605	.129	.0130698	.229	.0411871	.329	.0850124	.429	.144543	.529	.219787	.629	.310736	.729	.417393	.829	.539759	.929	.677832
.030	.0007069	.130	.0132732	.230	.0415476	.330	.0855300	.430	.145220	.530	.220618	.630	.311725	.730	.418539	.830	.541062	.930	.679292
.031	.0007548	.131	.0134782	.231	.0419097	.331	.0860492	.431	.145896	.531	.221452	.631	.312715	.731	.419687	.831	.542366	.931	.680754
.032	.0008042	.132	.0136848	.232	.0422733	.332	.0865699	.432	.146574	.532	.222287	.632	.313707	.732	.420836	.832	.543672	.932	.682217
.033	.0008553	.133	.0138929	.233	.0426385	.333	.0870922	.433	.147253	.533	.223123	.633	.314701	.733	.421986	.833	.544980	.933	.683682
.034	.0009079	.134	.0141026	.234	.0430053	.334	.0876160	.434	.147934	.534	.223961	.634	.315696	.734	.423138	.834	.546289	.934	.685148
.035	.0009621	.135	.0143139	.235	.0433737	.335	.0881415	.435	.148617	.535	.224801	.635	.316692	.735	.424292	.835	.547600	.935	.686616
.036	.0010179	.136	.0145267	.236	.0437436	.336	.0886685	.436	.149301	.536	.225642	.636	.317691	.736	.425448	.836	.548912	.936	.688085
.037	.0010752	.137	.0147411	.237	.0441151	.337	.0891970	.437	.149987	.537	.226485	.637	.318690	.737	.426604	.837	.550226	.937	.689556
.038	.0011341	.138	.0149571	.238	.0444881	.338	.0897272	.438	.150674	.538	.227329	.638	.319692	.738	.427763	.838	.551542	.938	.691029
.039	.0011946	.139	.0151747	.239	.0448628	.339	.0902589	.439	.151362	.539	.228175	.639	.320695	.739	.428923	.839	.552850	.939	.692503
.040	.0012566	.140	.0153938	.240	.0452390	.340	.0907922	.440	.152053	.540	.229022	.640	.321699	.740	.430085	.840	.554178	.940	.693979
.041	.0013203	.141	.0156145	.241	.0456168	.341	.0913270	.441	.152745	.541	.229871	.641	.322705	.741	.431248	.841	.555498	.941	.695456
.042	.0013854	.142	.0158368	.242	.0459961	.342	.0918635	.442	.153438	.542	.230722	.642	.323713	.742	.432412	.842	.556820	.942	.696935
.043	.0014522	.143	.0160606	.243	.0463770	.343	.0924011	.443	.154133	.543	.231574	.643	.324722	.743	.433579	.843	.558143	.943	.698416
.044	.0015205	.144	.0162860	.244	.0467595	.344	.0929401	.444	.154830	.544	.232428	.644	.325733	.744	.434747	.844	.559468	.944	.699898
.045	.0015904	.145	.0165130	.245	.0471436	.345	.0934822	.445	.155528	.545	.233283	.645	.326746	.745	.435916	.845	.560795	.945	.701381
.046	.0016619	.146	.0167415	.246	.0475292	.346	.0940249	.446	.156228	.546	.234140	.646	.327759	.746	.437087	.846	.562123	.946	.702867
.047	.0017349	.147	.0169717	.247	.0479164	.347	.0945692	.447	.156929	.547	.234998	.647	.328775	.747	.438260	.847	.563453	.947	.704353

APPENDIX J (Continued)
SOURCE INDEX

1. Visual Aids Service
   University of Illinois
   Division of University Extension
   1325 South Oak Street
   Champaign, Illinois 61820
2. Buehler Ltd.
   2120 Greenwood Street
   Evanston, Illinois 60204
3. Republic Steel Corporation
   Public Relations Dept.
   Republic Building
   Cleveland, Ohio 44101
4. Pickards-Mather and Co.
   Public Relations Dept.
   2000 Union Commerce Building
   Cleveland, Ohio 44115
5. Bureau of Mines
   United States Dept. of Interior
   Motion Pictures
   4800 Forbes Avenue
   Pittsburgh, Pennsylvania 15213
6. Reynolds Metal Co.
   Motion Picture Dept.
   2500 S. Third Street
   Louisville, Kentucky 40201
7. Association Films, Inc.
   Executive Offices
   600 Madison Avenue
   New York, New York 10022
8. Modern Talking Pictures Service
   621 North Skinker
   St. Louis, Missouri 63130
9. Indiana University
   Audio-Visual Center
   Bloomington, Indiana 47401
10. Coronet Films
    Sale Dept.
    Coronet Building
    Chicago, Illinois 60601
11. Phillips Petroleum Co.
    Chemical Dept.
    Bartlesville, Oklahoma 74003
12. The Society of the
    Plastics Industry, Inc.
    250 Park Avenue
    New York, New York 10017
13. University of Iowa
    Audio-Visual Center
    Division of Extension
    Iowa City, Iowa 52240
14. United World Films, Inc.
    221 Park Avenue
    New York, New York 10003
15. The Corning Museum of Glass
    Curator of Education
    Corning Glass Center
    Corning, New York 14830
16. A/V Communications Section
    Portland Cement Association
    Old Orchard Road
    Skokie, Illinois 60076
17. Tektronix, Inc.
    Film Library
    Delivery Station 50/420
    P. O. Box 500
    Beaverton, Oregon 97005
18. Quebec Government House
    The Film Officer
    Rockefeller Center
    17 West 50 Street
    New York, New York 10020
19. University of Missouri-Columbia
    Audio-Visual and
    Communication Service
    Whitten Hall
    Columbia, Missouri 65201
20. National Lubricating Grease Institute
    4635 Wyandotte Street, Room 202
    Kansas City, Missouri 64112
21. International Paper Co.
    220 East 42 Street
    New York, New York 10017

Pressing, Drawing, Extruding, and Rolling.

Fracture test. Breaking of a specimen and examining visually the fractured surface to determine such things as grain size and presence of defects.

G

Gage. The thickness or diameter used in connection with thin materials (usually not more than $\frac{1}{4}''$). The various standards are arbitrary and differ for ferrous and nonferrous materials and for sheets and wires.

Glass. An amorphous, noncrystalline solid made by fusing silica with a basic oxide.

Glass transition range. The temperature range through which liquid amorphous polymers become supercooled amorphous polymers and vice versa. As the liquid plastic cools, the motion of the polymer chains slows down. Below the glass temperature (Tg), chain motion is so limited that the structure appears to be a solid.

Grade. For wood, the designation of the quality of a log or of a processed piece of lumber.

Grain. An individual crystal in a metallic material; the quality, arrangement, direction, size, and appearance of the fibers in a piece of wood.

H

Hardness. The resistance of a material to localized plastic deformations.

Hardwood. Wood from a broad-leaved tree, characterized by the presence of vessels, as in oak, ash, or birch.

Heartwood. The inner part of the woody stem which, in the growing tree, no longer contains living cells. It is generally darker than sapwood, though the boundary is not always distinct.

Heat treatment. Heating and cooling a solid metal or alloy in such a way as to obtain desired conditions or properties. Heating for the sole purpose of hot-working is excluded from this definition.

Homopolymer. Polymer molecule formed from a single type of monomer or repeating structural unit.

Hooke's law. A law of mechanics which states that within the elastic zone of a material, the stresses are directly proportional to strains. This proportionality is constant and is known as the modulus of elasticity.

Hot-working. Deforming metal plastically at such a temperature and rate that strain hardening does not occur. The lowest such temperature is the recrystallization temperature.

I

Impact strength. The energy required to fracture a specimen under an impact (very rapidly applied) load.

Impact test. A test for determining the energy absorbed in breaking a specimen with a high-velocity, single-blow impact load.

Inert gases. The elements in the eighth row of the periodic table that have completed valence shells. These elements do not react chemically, and they exist as gases at room temperature.

Injection molding. Basically a plastic casting process similar to the die-casting process in metal work, in which semifluid plastic material is forced into a cold mold under pressure. Upon cooling, the mold opens to eject the finished casting.

Glossary

Intermetallic compound. A chemical compound of two or more elements which are constituents of an alloy system.

Ion. An atom or group of atoms exhibiting a net electronic charge by acquiring or losing valence electrons.

Ionic bond. Atoms bonded when valence electrons are interchanged to produce completed valence shells. Inorganic refractories are commonly bonded ionically.

Isomer. An organic molecule which has a fixed chemical composition but exhibits two or more structural arrangements of the component atoms.

K

Kaolinite. The prime ingredient of ordinary clay. It is a hydrated aluminum silicate.

Killed steel. Steel deoxidized to eliminate the reaction between carbon and oxygen during solidification.

Knitting. The process of making a fabric from a single thread or yarn. The basic structural pattern is a loop.

L

Lamella. A thin sheet of microfibrils forming a sublayer in a woody cell wall.

Laminate. A composite material usually in the form of a sheet or bar, composed of two or more layers so bonded that the composite material forms a structural member.

Latewood. The denser part of the growth ring. It is made up of wood cells having thicker walls, smaller radial diameters, and generally greater length than those formed earlier in the growing season; also called summerwood.

Lattice parameter. The distance between atoms or ions in a crystal.

Lattice structure. A regular, repetitive arrangement of atoms or ions.

Lignin. An amorphous substance which infiltrates and surrounds the cellulose strands in wood, binding them together to give a strong mechanical structure. It amounts to 15 to 30 percent by weight of the wood substance.

Linear polymer. A chainlike polymer composed of repeating molecules with very slight branching.

Lumber. The product of the saw- and planing-mill; wood not further processed than by sawing, resawing, standard planing, or cutting to length and width.

M

Macerate. To dissolve out the bonding material between plant cells to obtain separate, entire cells for examination.

Macromolecule. A molecule composed of many smaller molecules. Polymers are macromolecules.

Macrostructure. The structure of material that can be observed by unaided eye or low magnification.

Malleability. The characteristic of material that permits plastic deformation in compression without rupture.

Mechanical fastening. A linking process whereby materials are joined, permanently or semipermanently, with special locking devices such as screws, rivets, nails, bolts, keys, retainer rings, etc. The resultant joint is discontinuous.

Mechanical properties. Properties manifested in a material's reactions to applied forces and loads. Some mechanical properties are tensile strength, compressive strength, and fatigue strength.

Mer. The basic repeating structural unit of a polymer chain.

Metal. An opaque, lustrous, elemental substance that is a good conductor of heat and electricity and, when polished, a good reflector of light. Most elemental metals are malleable and ductile and are, in general, heavier than the other elemental substances. As to structure, metals may be distinguished from nonmetals by their atomic bonding and electron availability. Metallic atoms tend to lose electrons from the outer shells, the positive ions thus formed being held together by the electron gas produced by the separation. The ability of these free electrons to carry an electric current, along with the fact that the conducting power decreases as temperature increases, establishes one of the prime distinctions of a metallic solid.

Metallic bond. A bonding of metal atoms in which the atoms adopt a crystal structure and donate all valence electrons into an electron cloud.

Metallizing. A finishing process whereby a metal coating is sprayed onto the surface of a metal (or in some cases, plastic) part. The process is similar to electroplating and is also called thermal spraying or flame spraying.

Micelle. The crystalline portion of a bulk polymer structure produced when segments of a single chain are folded together or when neighboring chains are parallel so that the structural units of the chains adopt a repetitive pattern.

Microfibril. A bundle composed primarily of cellulose polymer chains which are regionally jointed to form crystallites; the smallest natural cell wall structural unit distinguishable with an electron microscope.

Microhardness. Hardness of microscopic parts (usually, the individual crystals or constituents) of material.

Microstructure. In metals, the structure revealed by examination of polished and etched specimens under the microscope.

Milling. A material-removal process employing a cylindrical tool having equally spaced, sharpened teeth on its surface, which moves through a fixed workpiece.

Modified wood. Wood processed to impart properties quite different from those of the original wood by chemical, resin, compressive, heat, or radiation treatments.

Modulus of elasticity. Within the elastic zone, the ratio of stress to corresponding strain. The modulus of elasticity is unique for each material.

Molecular architecture. The arrangement of the constituent atoms of polymers and, by extension, the intentional alteration or redesigning of macromolecules.

Molecule. A unit of matter, consisting of two or more chemically bonded atoms. The sum of the atomic weights is the molecular weight.

Monomer. Organic molecule capable of being converted into a polymer by chemical reaction with similar molecules or with other organic molecules.

N

Necking. Localized reduction of cross-sectional area of a tensile specimen under load.

Network polymers. The structure of thermosets. The individual polymer chains are bound together by chemical cross-linking and/or mechanical entanglements.

Neutron. A subatomic particle located in the nucleus of an atom. It has an arbitrary weight of one and is electrically neutral.

Glossary

Non-destructive testing. Testing methods that do not destroy the part to determine its suitability for use.

Nonferrous. Pertaining to metallic materials in which iron is not a principal element influencing the properties.

Normalizing. Heating a ferrous alloy to a suitable temperature range and then cooling in air to a temperature substantially below the transformation range (that is, the range within which the structure of the metal changes).

Notch sensitivity. A measure of the reduction in the load-carrying capacity of a material caused by the presence of stress concentration.

P

Paramagnetic. Pertaining to materials which exhibit a weak magnetic field of their own when subjected to a strong magnetic field.

Parenchyma. Tissue composed of wood cells which are usually shaped like tiny bricks, have simple pits, and frequently only a primary wall. Parenchyma functions mainly in the storage and distribution of food material.

Perforating. Piercing holes of desired shapes and arranged patterns in sheets, blanks, or formed parts.

Periodic table. A tabular arrangement of the elements by increasing atomic number which shows similarities in physical and chemical properties that can be explained by the electronic structure of the elements.

Permanent set. The percent of permanent deformation that remains after release of the stress that produced it.

Phase change. A change in a metal or alloy from one homogeneous, physically distinct substance to another. (In steel, a phase may be a metallic element, a compound, or a solid solution.)

Phosphorescence. Delayed reradiation of electromagnetic energy by excited electrons.

Photoconductivity. The ability of covalent compounds to conduct electricity when their valence electrons are excited by incident light on other electromagnetic radiations.

Photoemission. Excitation of electrons to leave a material by the application of electromagnetic radiation of the proper wavelength.

Physical properties. The properties associated with the physical characteristics of a material, such as density, electrical conductivity, and thermal expansion.

Pickling. Removing surface oxides from metals by chemical or elecrochemical reaction.

Piezoelectric effect. The reversible conversion of mechanical energy into electrical energy by complex metal oxide ceramic crystals.

Pit. In plant physiology, a recess in the secondary wall of a cell.

Pith. The central core of a woody stem, consisting mainly of parenchyma, or soft tissue.

Plaster. A special gypsum cement used as a building plaster and in the manufacture of wallboard. Plaster of paris and Keene's cement are closely related types.

Plastic deformation. An alternation of form or shape that remains permanently in a material after removal of the force that caused it.

Plasticity. The ability of a material to be nonelastically deformed without fracturing.

INDEX

Words defined in the glossary, pages 373–386, are italicized.

A

Abrading, 133, 134, 212
Abrasives, 207, 208
Abrasive saw, 132
ABS (acrylonitrile butadiene styrene), 166, 168, 169, 175
Absorptivity (optical), 18
Acetal, 166, 168, 169
Acetylene, 158
Acrylic, 166, 168, 169
Acrylonitrile butadiene styrene. See ABS
Addition polymerization, 159
Additives
  for petroleum products, 259, 263, 264, 273, 278
  for plastics, 160
Adhesion fastening, 149, 150, 192, 193, 213
Adhesives for wood, 234
Agglomeration of iron ore, 76
Air-carbon-arc cutting, 138
Air-entrained concrete, 210
Alkylation of petroleum, 273
Allotrope, 45, 56, 57
Allotropic changes, 45, 47
Alloying
  dislocation theory and, 64, 65
  elements for
    aluminum, 114
    cast iron, 84
    copper, 117
    magnesium, 121–123
    steel, 102
Alloys, 45, 47, 50–51, 325
Alloy steels, 99, 101–104
Alpha iron, 126, 128
Alumina, 110
Alumina hydrate, 110
Aluminate sheet, 196
Aluminum
  alloys, 112, 113, 131, 340–344
  annealing of, 130, 131
  crystal structure of, 44
  deoxidizer in steelmaking, 90
  general discussion of, 109–115
  oxide, 207, 208
  silicate, 209
  soap-based greases, 266

Alundum®, 207, 208
American Concrete Institute, 327
American Lumber Standards, 229, 230
American Society for Metals. See AMS
American Society for Testing and Materials. See ASTM
Amorphous plastics, 162, 164, 172–176
Amorphous regions of crystalline plastics, 177, 178
AMS classification of tool steel groups, 109
Anistropy and structural variations, 73
Annealing, 65, 129, 216
Annual ring, 218
Anodic metal, 17. See also electolytic cells
Anthracite coals, 271
Arc welding, 150, 151
Aromatic petroleums, 259
Aromatic polymers, 158
Asbestos, 58, 196
Ash of solid fuels, 270
Asphalt-based greases, 267
Asphalt concrete, 210
Asphalts, 254
ASTM
  classification of coals, 270
  requirements for diesel fuel oil, 336
  standard distillation test, 335
  standard jet evaporation test, 334
  standards, explanation of, 282
  standard specifications for fuel oils, 274
  standard water-and-sediment centrifuge test, 334, 335
  tentative specifications for gasoline, 277
Atactic pattern of polymers, 171
Atom, 28–32, 68
Atomic number, 28
Atomic theory, 27–33, 69. See also Bonding; Dislocation theory

Atomic weight, 28
Atomization of molten metal, 255
Austempering, 130
Austenite, 55, 127
Austenitic steels, 105
Aviation turbine fuels, 275, 276
Axial load, 303

B

Backbone of polymer chain, 170, 175
Baekelund, Leo Hendrik, 157
Bainite, 128
Bakelite, 181
Ball clay, 198
Band sawing, 131, 132
Barium-soap-based greases, 266
Bark, wood, 217
Bars, rolling of, 93
Basic oxygen steelmaking process, 85–87
Bauxite, 109, 208
BCC crystal lattice arrangement, 43, 44
BCC iron, 55, 126
Beam impact tests, 200
Bench tests for lubricants, 328
Bench-type tensile testing machines, 288
Bending
  metals, 138–140
  moment, 295
  stress, 224, 295
  tests, 292, 294–297
Benefication of iron ore, 76
Benzene, 158
Benzene ring, 170
Bessemer steelmaking process, 85, 88
Bifunctional monomer, 158, 165
Billets, 93–95
Bimetallic laminar composites, 253
Bipropellants, 278
Bisque firing, 195
Bituminous coals, 271
Blast furnace, 78, 79, 209, 268
Blister copper, 115, 117
Block molding, 215
Blooms, 92–94
Blow-and-blow process, 215

387

Index

Blowing glass, 215
Blow molding, 185
Board measure, 226
Body-centered cubic. *See* BCC
Boil test, 246, 248
Bonding
  ceramics, 56–58, 195
  chemical change and, 38
  of common industrial materials, 35
  effect of heat on, 71
  general discussion of, 33, 34
  of metals, 41–56
  of molecules, 34
  of organics, 58–60
  strength of, and physical states, 36, 37
  See also Covalent bond; Ionic bond; Metallic bond
Bonding (fabrics), 244
Bourdon tube for measuring loads, 290, 291
Box annealing, 129
Branched polymer, 171
Brass, 117–119
Brazing, 149
Breaking strength, 286. *See also* Fracture; Dislocation theory
Brick, 198–200, 296
Brinell hardness tests, 306–308
British thermal unit, 267
Brittle fractures, 283
Brittleness, 225. *See also* Fracture; Tensile testing
Bronze, 117, 119
Brucite, 121
Brushing metals, 153
Btu, 267
Buckling, 296
Butadiene, 158, 171, 172
Butane, 269
Butylene, 158

C

Calcium carbonate, 209. *See also* Limestone
Calcium-soap-based greases, 266
Calcium sulfate. *See* Gypsum
Calendaring plastics, 189
Calorie, 267
Calorimeter, 267, 333
Cambium layer in wood, 217
Capacitor, 16
Capped steel, 90
Carats, 124
Carbon
  allotropic changes of, 45
  basis for plastic, 158
  content of, in gray cast iron, 80

effects of, on steel, 97, 99, 129.
  fixed, 270
  See also Diamond; Graphite
Carbonaceous organic molecules, 158
Carbon-residue tests, 330, 331
Carbon steels, 96–99, 128
Carborundum®, 207
Carburizing, 71, 130
Cassiterite, 120
Casting
  ceramics, 212
  continuous, 93–96
  glass, 215
  methods of, 140, 141
  X-ray testing of, 321
Cast iron, 80–84, 297
Catalytic cracking, 273
Cathodic metal, 17. *See also* Electrolytic cells
Cellulose, 218, 241, 251. *See also* Organics
Cellulosics, 166
Cement, 208–210
Cementing of plastics, 192, 193
Cementite
  description of, 127
  elimination of, in nodular cast iron, 83
  formation of, in steel, 54, 55
  intermetallic compound, 51
  presence of, in gray cast iron, 81
  spheroidization of, 129, 130
Cement mortar, 210
Centrifugal casting, 141
Ceramics
  behavior of, under applied load, 60, 61
  bonding of, 56–68
  brittleness of, 67
  manufacturing processes for, 212, 215
  physical properties of, 345
  products made of, 199–211
  piezoelectric effect in, 72
  structure of, 195–197
Cermets, 254
Cetane number, 273, 335
Chain motion, 161, 170, 173–175
Chains
  alignment of, in crystalline polymers, 176, 177
  polymer, 158
  silicate, 58–60
Champleve metal enameling, 207
Characteristic properties, 14–19, 161–163, 178. *See also* Corrosion resistance;

Density; Electrical properties; Properties; Optical properties; Corrosion resistance
Characterization of plastics, 165. *See also* Characteristic properties; Chemical composition; Molecular architecture
Charcoal fuels, 269, 271
Charge. *See* Furnaces
Charpy impact test, 299, 300
Chemical blanking, 154
Chemical composition
  abrasives, 207, 208
  alloy steels, 101–104
  carbon steels, 98
  cements, 208, 209
  ceramics, 195–197
  copper alloys, 118, 119
  free-cutting carbon steels, 99
  gaseous fuels, 268
  glass, 205
  lubricants, 258–267
  magnesium alloys, 122
  organic fibrous materials, 241
  plastics, 158–160
  solid fuels, 271
  stainless steels, 104–106
  tool steels, 105, 107–108
Chemical elements, symbols and atomic weights of, 339
Chemical machining, 154
Chemical reactivity, 34, 36. *See also* Covalent bond; Ionic bond; Metallic bond
Chilled cast iron, 81, 82
China, 201
China clay, 197
Chlorofluorocarbon polymers, 262
Chrome tanning, 246
Chromium, 44, 104
Circuit (electrical), 68
Circular magnetization, 324
Circular sawing, 132
Classifications
  crankcase oil and transmission lubricants, 263
  fuel oils, 274
  greases, 329
  lead, 120
  wood, 226
  See also Designations
Clay
  effects of, in ceramic whiteware, 201
  slurry, 212
  types of, for ceramics, 197, 198

welding of, 213
Cleveland open-cup method of testing petroleum products, 332
Cloisonne metal enameling, 207
Close-die hydraulic pressing of composite materials, 256
Cloud-point test, 331
Coal, 45, 269, 270, 271
Coal gas, 268
Coated glasses, 204
Coefficient of thermal expansion, 15
Cohesive fastening, 150, 151, 213. 216. *See also* Adhesive fastening
Coke, 76, 77, 269, 271
Coke-oven gas, 268
Cold-heading, 142, 144
Cold-rolled steel, 90, 147
Cold-rolling mills, 147
Cold saw, 132
Cold-working, 63–65
Columbium, 125
Column-and-knee milling machine, 137
Combustion of petroleum products, 332. *See also* Ignition quality
Common brick, 199
Composite materials, 251–257
Compreg, 232
Compression molding, 184
Compressive strength
  definition of, 12
  formula for, 293
  measurement of, 162, 224, 292–294
  as result of metallic bond, 42
Concentration of iron ore, 76
Concrete
  different from cement, 209
  as particulate composite material, 254
  special types of, 210
  testing of, 294, 296, 326, 327
Condensation polymerization, 159
Condensor, 16
Conductivity, electrical. *See* Electrical conductivity
Conductivity, thermal. *See* Thermal conductivity
Conradson carbon-residue test, 330, 331
Consistency testing of lubricants, 328
Continuous casting, 93–96
Continuous filament, 253
Cooling curve of pure iron, 127
Copolymerization, 171

Copper
  alloys of, 117–120
  annealing of, 131
  crystal structure of, 44
  diamagnetism of, 71
  general discussion of, 115–120
Copper-corrosion test for petroleum products, 330
Copper-nickel alloy, 53
Corfam®, 250
Corium layer in hide, 246
Corrosion, 16–18
Corrosion resistance of plastics, 163, 164, 178
Cost of materials, 11, 155
Cotton, 242
Covalent bond
  electrical conductivity and, 67, 69
  electromagnetic radiation and, 72
  general discussion of, 38–40
  of glucose units in wood, 218
  in kaolinite, 196
  in organic fibers, 241
  in polymers, 59, 158
  of silicate tetrahedron, 196
  of silicon carbide, 207
  thermal conductivity in plastics and, 163
  See also Valence electrons
Cracked gasoline, 276
Cracking of petroleum distillates, 272, 273
Creep strength, 13, 162, 305
Cross-linking, 179, 180, 241
Crude petroleum, 258
Cryolite, 110
Crystalline plastics, 164, 176–178
Crystallites, 218
Crystal structure of metals, 41, 43–45. *See also* Grain, metal
Curie point, 70
Curing of plastics, 180, 184, 187, 188
Curl test in tanning, 246, 247
Cutting
  ceramics, 212
  glass, 215
  metals, 131–138
  plastics, 192
  wood, 233
Cyaniding, 130
Cyclic load, 12
Cyclohexane, 158

D

Dado joint, 234
Deflection in bending tests, 295, 296

Deflection temperature of plastics, 162
Deformation, 11, 61–66, 68, 282
  See also Visco-elasticity
Degree of crystallinity in plastics, 176
Delta iron, 126
Dendrite formation in solidifying metals, 46, 47
Density, 18, 19, 163, 222
Deoxidation in steelmaking, 90
Designation
  alloy steels (SAE-AISI), 96
  aluminum alloys (AA), 114, 115
  magnesium alloys (ASTM), 123, 124
  stainless steels (SAE-AISI), 106, 107
  tool steels (SAE-AISI), 107, 108
  See also Classification
Destructive distillation, 330, 331
Destructive testing, 281, 283–318
Diamagnetic materials, 70, 71
Diameters, areas of circles and, 366, 367
Diamond, 39, 40, 45, 56
Dibasic acid esters, 262
Die casting, 140, 141
Dielectric constant, 16, 224
Dielectric material, 224
Dielectric principle, 224
Dielectric strength, 16, 17, 163
Diesel fuels, 273, 274, 335–337
Diffuse-porous woods, 220
Diffusion and thermal energy, 71
Di-iso-octyl, 160
Dimer (organic unit), 158
Dislocation theory, 61–67
Dispersion hardening, 65
Displacement balance method for specific gravity, 333
Distillation of petroleum products, 259, 260
"Dividers method" of determining yield point, 286
Domains (magnetic), 70
Dolomite, 121
Dovetail joint, 234
Dowel joint, 234
Drawing metals, 145
Drilling metals, 135, 136
Drop-hammer forming, 142
Ductile cast iron, 82–84
Ductile fractures, 283
Ductility, 14, 286, 295–297. *See also* Dislocation theory; Tensile testing
Du Long's formula, 333

Index

Dy-Chek, 320
Dyeline, 320
Dye penetrant tests, 319, 320
Dynamic loads, 297, 298
Dynamic testing, 281, 316

E

Earlywood cells, 218
Ebonite, 181
Eddy current tests, 325, 326
Elastic behavior, 11
Elastic foams, 191
Elastic limit, 61
Elastic modulus, 60
Elasticity, 285, 287. *See also*
    Visco-elasticity
Elastomeric behavior, 173-175
Elastomers, 191
Electrical circuit, 68
Electrical conductivity, 16, 67, 68, 325
Electrical discharge machining, 153, 154
Electrical properties, 16, 223
Electrical resistance
    of conductor, 16
    in plastics, 162
    as reciprocal of conductivity, 67
    temperature and, 68
    variations of, in wood, 223, 224
Electric
    arc, 151
    current, 67
    steelmaking process, 85, 87, 88
Electrochemical corrosion, 17, 18
Electrochemical machining, 154
Electrolyte, 17
Electrolytic cells, 17, 116, 117
Electrolytic decomposition in production of metal powders, 255
Electrolytic tough pitch copper, 117
Electromagnetic radiation, 71, 72
Electron
    effect of electromagnetic radiation on, 72
    function in electrical circuit, 68
    negative charge of, 28
    as subatomic particle, 27. *See also* Atomic theory; Covalent bond; Ionic bond; Metallic bond; Valence electrons
Electron acceptor atom, 40
Electron cloud, 42, 71
Electron donor atom, 40
Electron-neutron radiography, 323

Electron shell theory, 29-33
Electrothermic process, 121
Emulsion polymerization, 160
Endurance limit, 302, 304
Energy absorption, 298-302
Energy, thermal, and diffusion, 71
Engine lathe, 137, 138
Environment, 10, 11, 164
Epidermis in hide, 245
Epoxy, 167-169, 180
Epoxy resins, 59
Epoxy thermosets, 181
Equilibrium diagrams, 52, 53
Esters, 262
Ethers, 262
Ethylene, 158
Eutectic mixture, 53, 54, 128
Eutectoid mixture, 55, 56, 128
Evaporation of petroleum products, 332
Existent-gum tests, 334
Explosive forming, 153, 154
Extenders (plastics), 160
Extensometer, 286
Extruding, 145-147, 182-184, 212

F

Fabrics, 241
Face brick, 200
Face-centered cubic. *See* FCC
Failure of materials, 282, 283, 296
Fastening
    adhesion, 149, 150, 192, 193, 213
    ceramics, 213
    metals, 131, 148-151
    plastic parts, 192
    wood, 234
Fatigue, 12, 302-304
Fatty oils, 261
FCC
    crystal lattice arrangement, 43, 44
    iron, 54
    region, 65
    structure of copper-nickel alloy, 53
Feldspar, 196, 198, 201
Felting, 244
Ferrimagnetism, 69
Ferrite, 55, 127, 128
Ferritic malleable cast iron, 84
Ferritic steels, 105
Ferromagnetism, 69, 70
Ferrosilicon, 90
Ferrous metallic materials, 75-107, 126-130.

See also Iron; Steel
Fiberboard, 231
Fiberglass-reinforced plastic films, 252
Fibers, 241-244, 251-253
Fiber saturation point, 225
Fibrils
    in hide, 245
    in woody cells, 218
Fibrous ceramics, 58
Fibrous composite materials, 251-253
Filament, continuous, 253
Filaments in fibrous materials, 241
File test, 317
Fillers (plastics), 160
Finishing
    ceramics, 213
    manufactured textiles, 244
    metals, 131, 151-153, 206, 207
    plastic products, 193
    wood, 234
Fire brick, 202
Fireclay, 198
Fire point test, 331, 332
Fire-refined tough pitch copper, 117
Firing ceramics, 195, 196
Fixed-bed milling machine, 137
Fixed carbon, 270
Fixed oils, 261
"Flaking of the scale" method of determining yield point, 286
Flame hardening, 130
Flame spraying, 152
Flashing in brick manufacture, 199
Flash point test, 331, 332
Flat drawing of glass, 215
Flexible foams, 190, 191
Flexural
    load, 303
    strength tests, 162
Flint clay, 198
Flow in solids, 13
Flow temperature, 162
Fluorescence, 72
Fluoroscopy, 323
Foams (plastic), 190, 191
Forging, 141-144
Formability of industrial materials, 11, 12
Forming
    ceramics, 212
    metals, 131, 138-148, 153
    plastics, 181-186
    wood, 233
Formulas
    addition polymerization, 159

Index

bending stress, 295
board feet, 226
breaking strength, 286
Brinell hardness number, 306
compressive strength, 293
condensation polymerization, 159
conductivity, 67
density of wood, 222
Dulong's, 333
elongation, 286
hardness number, 306
heating value, 333
metric conversion, 347–349, 351
moisture content of wood, 221
reduction of area, 287
resistance, 16, 67
resistivity, 16
specific gravity in wood, 222
thermal conductivity of wood, 223
thermal resistance of wood, 223
ultimate tensile strength, 286
Vickers hardness number, 315
Fosterite, 57
Fracture
  dislocation theory and, 66, 67
  fatigue and, 303
  in impact tests, 298, 299
  modulus of, in bending tests, 295
  as response of material to load, 11
  resistance of ductile metals to, 61
  types of, 283
Fracture strength, 286
Free bend test, 297
Free-cutting carbon steels, 97, 99
Friability of solid fuel, 269
Friction, 258, 261
Fuels, 254, 255, 265–278, 332–337
Furnaces
  air, 80
  basic oxygen, 86–88
  blast, 76–79, 115, 120
  cupola, 80
  electric arc, 87
  electrolyte, 110
  open-hearth, 80, 85, 86, 88
  regenerative, 85
  reverbatory, 85, 115–117, 120
Furniture, 238, 239

G

Galena, 120
Galvanic series, 18
Gamma iron, 126, 127
Gamma ray, 321, 322
Gas chromatography, 333
Gaseous fuels, 267–269, 332, 333
Gases, 36, 37, 72
Gasolines, 163, 276, 277
Gas-tungsten-arc cutting, 138
Gas turbine fuels, 275, 276
Gas welding, 150
Glacial clay, 198
Glass
  behavior under applied load, 60
  general discussion of, 203–206
  manufacturing processes, 215–216
  physical properties of, 359
  as supercooled liquid, 58, 61
Glass transition temperature of polymers, 173–175, 177, 180, 181
Glazing, 213, 214
Glost firing, 195
Gold, 44, 124
Goodyear, Charles, 157
Graft polymerization, 171, 172
Grain
  hide, 245, 246
  metal, 47–50, 65
  wood, 222
Graphite, 45, 46, 80, 81, 265
Gray cast iron, 80, 82
Greases, 163, 263–265, 329
Grinding, 133
Grinding wheels, 208
Grog (ceramic), 198, 202
Guided bend test, 297
Guillotine shears, 132, 133
Gypsum, 209, 211

H

Hacksawing, 131, 132
"Halt of the gage" method of determining yield point, 286
Hand forming of ceramics, 212
Hard magnets, 70
Hardness, 14, 306
Hardness tests, 162, 225, 292, 305–318, 325
Hardwoods, 219, 220, 226–228, 360–363
HCP crystal lattice arrangement, 43
Heartwood, 218
Heat (furnace load), 86. *See also* Furnaces
Heat conductivity, 69, 71. *See also* Thermal conductivity
Heating value, 267, 276, 332, 333
Heat of combustion, 335
Heat treatment, 65, 66, 125–131
Hematite, 75
Hemp, 242
Hexagonal close-packed crystal lattice arrangement, 43
Hides, 244
High alloy steels, 103–109
High-carbon steels, 98, 99, 101, 129, 130
Hollow-spindle lathe, 138
Home construction, 237, 238
Hooke's law of elasticity, 285, 287
Hot-rolled steel, 90–95
Hot working, 67. *See also* Forging and Hot-rolled steel
Hyatt, John Wesley, 157
Hydraulic cement, 208
Hydraulic grip, 290
Hydraulic tensile testing machine, 288–292
Hydrophilic plastics, 162, 163
Hypereutectoid steels, 129

I

Ignition quality of fuel, 273, 276, 335
Impact strength, 13, 162, 297–302
Impreg, 232
Impregnation of skeletal composite, 257
Induction hardening, 130
Industrial materials
  behavior of, 60–73
  selection of, 9–25, 155
  structure of, 45–60
  testing of, 281–338
  See also Ceramics; Composite materials; Fibers; Fuels; Glass; Lubricants; Metals; Natural leather; Plastics; Structural clay products; Synthetic leathers; Synthetic lubricants; Wood
Industrial materials technology, definition of, 9
Inert gases, 36, 37
Infiltration of skeletal composite, 257
Injection molding, 181, 182
Inorganic fibers, 241
Inorganic polymers, 58, 195, 196
Inspection, 283. *See also* Nondestructive testing
Insulating abilities of plastics, 163
Intergranular fracture, 283
Intermetallic compounds, 51, 52, 127
Interstitial solutions, 51, 54, 55
Investment casting, 141

Index

Ionic bond
  in ceramics, 195
  electrical conductivity and, 67, 69
  general discussion of, 40, 41
  in kaolinite, 196
Ionic conduction, 162, 163
Ions, 40, 42, 67–69
Iron, 33, 44, 45, 55. *See also* Ferrous metallic materials
Iron carbide. *See* Cementite
Iron-carbon equilibrium diagram, 55
Iron, cast, 80–84
Iron ore, 75, 76
Island silicate structure, 57
Isobutylene, 158
Isomerism, 170
Isotactic chains, 170, 171, 176
Isotope, 29
Izod impact test, 299, 300

J

Jasper, 76
Jet fuels, 275, 276
Jiggering of ceramics, 212, 214
Joints in wood, 234
Jute, 242

K

Kaolin, 197
Kaolinite, 196, 197
Keene's (Keenan's) cement, 211
Killed steel, 90
Kinematic method of viscosity testing, 328, 330
Knitting, 244
Knocking index of gasoline, 276

L

Lake copper, 117
Lamella of woody cells, 218, 219
Lamellar structure of pearlite, 128
Laminar composite materials, 253
Laminating, 186–188, 233
Laser welding, 154, 155
Latewood cells, 218
Latex, 160
Lathes, 135, 137, 148
Lattice parameter of chemical elements, 43
L/D ratios for compression test specimens, 293, 294
Lead, 120, 265
Lead carbonate, 160
Lead-tin alloy, 53, 54
Leaded steels, 104

Leather, 244–251
Ledeburite, 128
Length, metric conversion chart for, 352, 353
Lignin in woody cells, 219, 232, 251
Lime, 129
Lime mortar, 210
Limestone, 76, 77, 86, 87
Limoge metal enameling, 207
Limonite, 75
Liquefied petroleum gases, 268, 269
Liquid fuels, 272–277, 334–337
Liquid penetrant tests, 319, 320
Liquid slip, 206, 207
Linear homopolymer, 165
Linen, 242
Lithium-soap-based greases, 266
Load
  axial, 303
  in bending tests, 295
  cyclic, 12
  in fatigue testing, 302, 303
  flexural, 303
  integrity of shape and, 10
  measurement of, 290–292
  in microhardness test, 316
  resistance of materials to, 60, 61, 297, 298
  in Rockwell hardness test, 308
  strength of material and, 10
  in tensile testing, 284, 290
  torsional, 303
  in Vickers hardness test, 315
Longitudinal magnetization, 324
Lost wax casting, 141
Low-alloy steels, 99, 101
Low-carbon steels, 97, 98, 101
Lubricants, 258–267, 327–332
Lucite, 176

M

Machining metals, 131–138, 153, 154
Machining in production of metal powders, 255
Macromolecules, 158, 163, 165. *See also* Chains
Magnesite, 121
Magnesium, 83, 121–124
Magnesium alloys, 121–123
Magnesium oxide, 57
Magnetic analysis tests, 323–325
Magnetism, 69–72
Magnetite, 75
Magnetization, 324
Malleable cast iron, 84
Manganese steels, 104

Manufacturing processes. *See* Abrading; Bending; Blowing; Blow molding; Bonding (fabrics); Casting; Cutting; Drawing; Drilling; Electrochemical machining; Electrical discharge machining; Explosive forming; Extruding; Fastening; Finishing; Flat drawing; Forging; Forming; Glazing; Grinding; Knitting; Laminating; Molding; Perforating; Planing; Plasticization; Pressing; Rolling; Sawing; Shaping; Shearing; Slip-casting; Slip cementing; Solvent cementing; Spinning; Tube drawing; Turning; Weaving; Welding
Manufacturing requirements, 9, 10
Martempering, 130
Martensite, 128
Martensitic steels, 105
Materials
  failure of, 282, 283, 296
  specifications for, 24. *See also* Industrial materials
Matrices in composite materials, 251–254
Matte (impure copper), 115
Maximum useful temperature of plastics, 162
Measurement (metric system), 346–359
Mechanical crystallization of linear plastics, 176, 177
Mechanical energy, effects of, 72
Mechanical fastening of metals, 148, 149
Mechanical grips for tensile testing machines, 290
Mechanical properties
  amorphous plastics, 173
  crystalline plastics, 177, 178
  general discussion of, 11–14
  network plastics, 180, 181
  plastics, 161, 162
  semicrystalline plastics, 178, 179
  wood, 224
  wrought aluminum alloys, 340–344.
  See also Compressive strength; Creep strength; Ductility; Hardness; Fatigue; Impact strength; Properties; Tensile strength

Index

Mechanical stretch forming of plastics, 186
Mechanical testing of plastics, 162. *See also* Compression tests; Hardness tests; Impact tests; Tensile tests
Medium-alloy steels, 101, 103
Medium-carbon steels, 97–99, 101
Melamine, 167, 181
Mercerization, 244
Metal ores
  bauxite, 109
  cassiterite, 120
  galena, 120
  hematite, 75
  limonite, 75
  magnetite, 75
  magnesium chloride, 121
  siderite, 75
Metals
  basis for classification of, 44
  behavior under applied load, 60, 61
  bending tests for, 296, 297
  chemical definition of, 36
  crystal structure of, 43–45. *See also* Grain, metal
  electrical conductivity of, 67, 68
  enameling of, 206, 207
  ferrous, 75–107, 126–130
  nonferrous, 109–125, 130–131
  processing of, 131–155, 255–257
  softening of, 65, 66
  strengthening of, 62–65
  thermal conductivity of, 71
  thermoelectric effect on, 68, 69
Metallic bond, 41–43. *See also* Crystal structure of metals; Dislocation theory
Metallic lattice, 43
Metallic soap, 263
Metallizing, 152, 153
Met-L-Chek, 320
Metric system, 346–359
Micelle crystallization of linear plastics, 176
Microfibril in woody cell, 218
Microgel greases, 267
Microhardness tests, 307, 316
Mill products
  aluminum, 110, 111
  copper, 117
  magnesium alloys, 123
  steel, 88–94
  tin, 121
  zinc, 121
Milling (metals), 136, 137, 225
Mining of iron ore, 75, 76
Mixed-base petroleums, 259

Mixtures in alloys. *See* Eutectic mixture; Eutectoid mixture
Modulus of elasticity
  definition of, 224, 225
  determination of, in bending tests, 295, 296
  determination of, in eddy current tests, 325
  specimen for testing, 293
Modulus of fracture, 295
Mohs scale of hardness, 317, 318
Moisture content of wood, 221, 225
Molding, 181, 184, 185, 189, 215
Molecular architecture of plastics, 165, 170–172, 177, 178
Molecular packing, 165
Molecular weight of plastics, 172, 175
Molecule, diatomic, 38, 40
Molecules, bonding of, 34
Molybdenum disulfide, 265
Molybdenum steels, 104
Monel, 53
Monobasic acid esters, 262
Monomers, 158, 164, 180
Monopropellants, 278
Mortar, 210
Mortise-and-tenon joint, 234
Multidirectional magnetization, 324
Multiple-spindle drill presses, 135, 136
Multi-soap based greases, 266

N

Naphthene-base petroleums, 258, 259
Napthene polymers, 158
Natural fibers, 241, 242
Natural gases, 268
Natural leather, 246–249
Network plastics, 158, 164, 165, 179–181
Neutralization numbers of lubricating oils, 332
Neutron, 27, 28
Nickel-aluminum, 51
Nickel steels, 103, 104
Nitriding, 130
Nodular cast iron, 82–84
Nondestructive testing, 281, 318–326
Nonferrous metallic materials, 75, 109–125, 130, 131
Nonhydraulic cement, 208

Non-linear homopolymers, 171
Nonmetals, chemical definition of, 36
Nonporous woods, 219, 220
Normalizing, 129, 130
Notch in impact specimen, 298, 299
Nuclear fuels, 278
Nucleus
  of atom, 28
  of grain (metals), 47. *See also* Dendrite formation
Numerically controlled lathe, 138
Nylon, 166, 168, 169

O

Octane scale, 276, 277
Oils, fixed. 261
Olefin, 158
Open-hearth process, 85, 86
Open-pit mining, 75, 76
Optical properties, 18, 176
Optical translucency of crystalline plastics, 178
Ore. *See* Metal ores
Organic covalent bond, 38, 39
Organic fibrous materials, 241
Organics, 58–60, 72. *See also* Leather; Plastics; Wood
Oxidation, 16
Oxygen-bomb calorimeter test, 335
Oxygen cutting, 138
Oxygen lance, 86
Ozone in degradation of rubber, 17

P

Paraffin-based petroleums, 258
Paramagnetic materials, 70
Parenchyma (woody cells), 219
Parkes process for production of lead, 120
Particulate composite materials, 253–257
Pearlite, 55, 56 128, 129
Pearlitic malleable cast iron, 84
Peat, 269, 271
PEG process, 232
Pendant group in polymer molecules, 170
Penetrators for hardness testing, 313, 315, 316, 328
Penetrometer, 328
Percent of elongation, 286
Percent of reduction of area, 286, 287, 364, 365

393

Index

Perforating (metals), 133
Periclase (magnesium oxide), 57
Periodic table of chemical elements, 29, 30, 33
Permanent magnets, 70
Permeability of leather, 248, 249
Petrochemicals, 158. *See also* Gaseous fuels; Liquid fuels; Petroleum lubricants; Plastics
Petroleum, 259, 260. *See also* Gaseous fuels; Liquid fuels; Petroleum lubricants
Petroleum lubricants, 259–265
Phase transformation of metals, 47, 48, 71
Phenol formaldehyde, 181
Phenolic, 167–169
Phenolic plastics, 157
Phloem in wood, 217
Phosphate esters, 262
Phosphorescence, 72
Photoconductivity, 72
Photoemission, 72
Photosensitivity, 124
Physical properties
 ceramic materials, 345
 glasses, 359
 wood, 222
 See also Characteristic properties; Mechanical properties
Physical states of matter, 36
Piezoelectric effect, 72
Piezoelectric crystal, 323
Pig iron, 76–80
Pigs, cast iron, 79
Pipe solidification in steel, 90
Pith, 218
Plainsawed boards, 222
Planing (metals), 133, 134
Plasma-arc cutting, 138
Plasters, 210, 211
Plastic coating, 190
Plastic deformation. See Deformation
Plastic foams, 190, 191
Plasticization of wood, 232
Plasticizers, 160
Plastic processes and products, 181–193
Plastics
 amorphous, 162, 164, 172–176
 classification of, 164, 166–169
 crystalline, 164, 176–178
 definition of, 158
 manufacture of, 158–160
 mechanical testing of, 162
 properties of, 160–164
 specimens of, for tension impact test, 298

structure of, 164, 165, 168–181
 See also Polymers
Platelets in kaolinite, 196
Plate steel, 93
Platinum, 124
Plexiglas, 176
Plunger method of injection molding, 182
Plywood, 229
Pneumatic grip for tensile testing machines, 290
Polycarbonate, 166, 168, 169, 175
Polyethylene
 effect of semicrystallinity on, 178, 179
 as linear plastic, 165
 molecular architecture of, 170
 processing and composition of, 167
 structure and properties of, 58, 59, 168, 169, 171, 175
Polyethylene adipate, 159
Polyethylene glycol process, 232
Polyfunctional monomers, 158, 180
Polyglycol ethers, 262
Polymeric materials, 17. *See also* Plastics; Polymers
Polymerization, 158–160, 172, 231, 273. *See also* Copolymerization; Graft polymerization
Polymers
 behavior under applied load, 60, 61
 definition of, 58
 effect of electromagnetic radiation on, 72, 160, 163
 effect of heat on, 71
 in lubricants, 262
 See also Plastics
Polymorph. *See* Allotrope
Polypropylene, 167, 168, 169, 175
Polystyrene
 addition polymerization of, 159
 copolymerization of, 171, 172
 cross-linking of, 180
 as linear plastic, 165
 processing and composition of, 167
 structure and properties of, 168, 169, 170, 175
 susceptibility of, to corrosion, 163
Polyvinyl chloride, 167, 175
Porcelain, 201–203
Porcelain enamels, 206, 207
Porous woods, 219, 220
Portland cement, 208, 209
Potter's wheel, 212

Pour-point test for oil, 331
Powder cutting, 138
Powder metallurgy, 255–257
Precast concrete, 210
Precious metals, 124, 125
Precipitation numbers of lubricating oils, 332
Press brake bending, 140
Pressing, 142, 144, 145, 212, 215
Prestressed concrete, 210
Pressure forming plastics, 186
Producer gases, 268
Production drill presses, 135
Propane, 269
Properties
 aluminum, 111, 114
 carbon steel, 97, 100
 copper, 117
 definition of, 11
 determination of, by tensile tests, 284
 effect of structural variations on, 72, 73
 gaseous fuels, 268, 269
 glass, 205, 359
 gold, 124
 gray cast iron, 80
 lead, 120
 magnesium and its alloys, 121–123
 malleable cast iron, 84
 natural leather, 248, 249
 nodular cast iron, 83
 pig iron, 79, 80
 plastics, 161–163
 platinum, 124
 silver, 124
 solid fuels, 269–271
 structural variations and, 72, 73
 tin, 121
 white cast iron, 81
 wood, 220–226
 zinc, 121
 See also Characteristic properties; Mechanical properties; Physical properties
Proportional limit, 225, 285
Propylene, 158
Proteinic fibers, 241, 242, 245
Proton, 27, 28
Proximate analysis, 333
Pug mill in ceramics manufacture, 199
PV, 248, 249
Pvc acetate copolymer, 167
Pyrometallurgical process, 121
Pyrometer, 126
Pyrometric cone, 197, 198

Index

Q

Quantum mechanics, 29
Quartersawed boards, 222

R

Rabbet joint, 234
Radiation. *See* Electromagnetic radiation; X-ray as source of radiation
Radiographic tests, 320-323
Random mat fiber arrangement, 252
Raw materials for plastics, 158
Rebound hardness tests, 306, 316, 317
Reciprocating screw method of injection molding, 181, 182
Recrystallization of metals, 50, 65, 129
Reduction of area gage in tensile testing machines, 292
Reduction in production of metal powders, 255
Reflectance, 18
Refractivity, 18
Refractories, 202, 203
Refractory insulators, 202, 203
Refractory metals, 125
Reinforced concrete, 210
Reinforced plastics, 188, 189
Reinforcing agents (plastics), 160
Resin ducts in softwoods, 220
Resins, 158, 164
Resistance, electrical. *See* Electrical resistance
Resistance welding, 150, 151
Resistivity of insulators, 16
Rigid foams, 190
Rimmed steel, 90
Ring-porous woods, 220
Rocket propellants, 278
Rockwell hardness tests, 307-309, 313-315
Rods, rolling of, 93
Roll bending, 140
Rolling
  glass, 215
  metals, 90-93, 147
Rotating-beam machine, 303
Rotational molding, 189
Rubber, 58, 59, 157, 163

S

SAE viscosity classification of oils, 263
SAE-AISI classification of tool steels, 107
SAE-AISI system of steel designation, 94, 96
Safety, 10, 322
Salt glaze, 200
Salt solution and ionization, 69
Sand casting, 140, 141
Sanforizing, 244
Saponification numbers, 332
Sapwood, 217
Sawing, 131, 132, 212, 215
Saybolt method of viscosity testing, 328
Scleroscope hardness test, 307
Scratch hardness tests, 306, 317, 318
Second-phase particles and dislocation theory, 64, 65
Segregation of ingredients in steel, 90
Selection of materials, 9-25, 155
Selenium, 69
Semicrystalline plastics, 165, 178, 179
Semi-guided bend test, 297
Semikilled steel, 90
Semivitreous clay products, 201
Sensitive drill presses, 135
Service life, 10, 14
Service requirements, 10, 11, 20, 21
Shaft mining, 75
Shale, 198
Shaping (metals), 133-135
Shearing, 132, 133, 212, 215, 296
Shear stress in wood, 224
Shear tests, 292, 305
Sheet blow molding, 186
Sheet steel, 93
Sheet-steel enameling process, 206, 207
Shielded-metal-arc welding, 150, 151
Shore Scleroscope, 317
Shotting in production of metal powders, 255
Shrinkage in wood, 221
Siderite, 75
Silica, 201, 203
Silicate chains, 58, 195, 196
Silicate esters, 262
Silicate sheets, 196
Silicate tetrahedron, 57, 58, 195, 196
Silicate unit. *See* Silicate tetrahedron
Silicon carbide, 207
Silicon in gray cast iron, 80
Silicon polymers, 262
Silicons as solid lubricants, 265
Silk, 242

Silver, 44, 70, 124
Singeing fabrics, 244
Sintering of metal-powder composites, 256
Sisal, 242
Skeletal particulate composites, 256, 257
Skins in leather processing, 244
Slabbing mill, 93
Slabs (steel), 93, 94
Slag, 78, 79, 87
Slip (clay slurry), 212
Slip clay, 198
Slip-casting of ceramics, 212, 213
Slip cementing of ceramics, 213, 214
Slip mechanism in structure of materials, 61, 62, 66, 67, 196, 283, 303
Slump test for concrete, 210, 326, 327
S-N diagram, 304
Soaking pit for steel, 90
Sodium aluminate, 110
Sodium chloride, solution and ionization of, 69
Sodium-soap-based greases, 266
Soft magnets, 70
Softwoods, 219, 220, 226, 227, 229, 230, 360-363
Soil cement, 210
Soldering, 150
Solid fuels, 254, 255, 269, 270, 333
Solid lubricants, 265
Solidification
  of metals, 47-50, 52-56, 126-131
  of plastics, 173, 176
Solid solutions in metals, 50-56, 127, 128
Solution heat treating for substitutional and interstitial alloys, 66
Solvent cementing of plastics, 192, 193
Solvents of plastic, 163
Sound waves, 72
Space lattice structure of crystalline materials, 42-45, 47-56, 126-129
Special glasses, 204, 206
Specific gravity, 19, 221, 222, 225, 332, 333
Specific heat, 15
Specimens
  bending tests, 295-297
  compression testing, 293, 294
  fatigue testing, 303
  impact testing, 298-300

395

Index

magnetization of, 324
microhardness testing, 316
tensile testing, 287
Vickers hardness test, 316
Spectrometry, 333
Spheroidization of steel, 129, 130
Spherulitic graphite cast iron, 82–84
Spinning
 fibers, 243
 metals, 148
Spline edge joint, 234
Split molding of glass, 215
Spontaneous ignition temperature, 273
Spotcheck, 320
Stabilizers (plastics), 160
Stainless steels, 104–106
Staple fibers, 241
Static loads, 297, 298
Static testing, 281
Stationary-die bending machine, 140
Staypak, 232
Steel, 84–109
Steelmaking processes, 84–88
Stiffness-in-flexure tests, 162
Stoneware, 200
Straight run gasoline, 276
Strain, 284, 291. *See also* Tensile strength; Tensile testing
Strain gage, 286
Strength, 283
Strength properties of woods, 360–363
Strength-to-weight ratios of fibrous composite materials, 251
Stress, 284. *See also* Tensile strength; Tensile testing
Stress relieving, 65. *See also* Heat treatment
Stress-rupture failure, 14
Stress-strain curve, 284, 285
Stretch-draw forming, 144
Structural clay products, 199, 200
Structural variations and properties, 72, 73
Structure of materials. *See* Atomic theory; Bonding; Chains; Chemical composition; Crystal structure of metals; Fibers; Grain; Molecular architecture of plastics; Periodic table of chemical elements; Physical states of matter; Space lattice structure of crystalline materials

Styrene, 171, 172
Styrene butadiene, 175
Styrofoam, 190
Subatomic particles, 27, 28. *See also* Electron
Substitutional solid solution of alloys, 51
Sulfuric acid, 69
Super alloys, 125
Supercooled liquid, 173
Supersaturated solid solution, 128
Surface defects, 318
Surface finish, 303
Surface grinding, 133
Surface hardening processes, 130
Suspension polymerization, 160
Swelling in plastics, 163
Syndiotactic pattern, 171
Syndiotactic chains, 176
Synthetic leathers, 249–252
Synthetic lubricants, 260–262
Synthetic rubber, 171, 172

T

Taconite, 76
Tacticity, 165, 170
Tanning of natural leathers, 246–248
Tannins, 246
Tantalum carbide, 208
Tellurium, 69
Temperature, effect of
 on amorphous plastics, 173
 on crystalline plastics, 177
 on electrical conductivity, 68
 in impact tests, 299, 300
 on linear plastics, 170
 on magnetism, 70
 on thermosets, 181
 on visco-elasticity, 161, 162
 on wood, 226
 See also Thermal properties
Temperatures, metric conversion chart for, 354–358
Tempering, 65, 130, 216
Tensile strength
 definition of, 11
 different from impact strength, 13
 measurement of, 286
 metallic bond and, 42
 of natural leather, 249
Tensile stress in wood, 224
Tensile testing
 cause of failure in, 62
 general discussion of, 284–292
 machines for, 287–292
 percentage reduction of area table for, 364, 365

plastics, 162
sample data recording sheet for, 368
Tension impact tests, 299, 300
Testing. *See* Bending tests; Compression tests; Concrete; Destructive testing; Electric analysis test; Failure of materials; Fatigue tests; Fuels; Hardness tests; Impact tests; Lubricants; Magnetic analysis test; Nondestructive testing; Radiographic tests; Tensile tests; Ultrasonic tests
Texture in wood, 222
Tg, 173, 177
Thermal conductivity
 definition of, 15
 of metals, 71
 of plastics, 163
 of wood, 222, 223
Thermal cracking, 273
Thermal cutting, 138
Thermal energy, 71, 72, 170
Thermal expansion in wood, 223
Thermal properties, 14–16
Thermal resistance in wood, 223
Thermal spraying, 152
Thermal stresses, 15
Thermionic emission, 71, 72
Thermocouple, 69, 126
Thermoelectric effect on metals, 68, 69
Thermoforming plastics, 185, 186
Thermoplastics
 calendaring of, 189
 definition of, 164
 injection molding of, 181, 182
 table of, 166
 thermoforming of, 185, 186
 See also Amorphous plastics; Crystalline plastics; Semicrystalline plastics
Thermosets
 approximate yield point of, 162
 compression molding of, 184
 cross-linked structure of, 179–181
 definition of, 164
 injection molding of, 182
 table of, 167
 transfer molding of, 184, 185
 use of, in high-pressure laminating, 186-188
 use of, in reinforced plastics, 188, 189. *See also* Network plastics
Thermosetting resin, 232
Tin, 117, 120, 121

Index

Titanium, 125
Titanium alloys, 131
Titanium steels, 104
Toluene, 158
Tool steels, 105–107
Torsional load in fatigue testing, 303
Torsion impact tests, 300
Torsion tests, 305
Toughness
  description of, 225
  different from ductility, 14
  as impact resistance, 13
  measurement of, 298
Tracer lathe, 138
Tracheids (woody cells), 219
Transcrystalline fracture, 283
Transducer, 72
Transfer molding, 184
Transition metals, 36
Translucency in china, 201
Transverse flexure tests. *See* Bending tests
Tube drawing glass, 215
Tubing (metal), 140
Tungsten, 125
Tungsten carbide, 208
Turning, 136–138, 212, 213
Twist drill, 135

U

Ultimate analysis, 333
Ultimate tensile strength, 286
Ultrasonic tests, 323
Uniaxial compression loading, 283, 284
Uniaxial tensile loading, 283, 284
Unit cell in crystal lattice, 43, 44
Universal fatigue testing machine, 302, 303
Universal impact testing machine, 300, 301
Universal testing machine, 288, 289, 293, 295
Urethane, 167

V

Vacuum degassing, 89
Vacuum forming plastics, 186
Valence of metals, 44
Valence electrons
  effect of electromagnetic radiation on, 72
  and electrical conductivity, 67
  and magnetism, 69, 70
  and resistance in plastics, 162
  in silicate tetrahedron, 196
  thermionic emission of, 71, 72
Valence shell and chemical reactivity, 34, 36
Vanadium, 125
Vanadium steels, 104
Van der Waals bond, 34, 40
Van der Waals forces
  in amorphous plastics, 173
  in cellulose polymers, 218
  general discussion of, 37, 38
  in kaolinite, 196
  in organic fibers, 241
  in polymers, 58
Vapor condensation in production of metal powders, 255
Vegetable tanning, 246
Verification in hardness testing, 315, 316
Vessels in wood, 219, 220
Vibrational amplitude, 68
Vicat softening point, 162
Vickers hardness test, 307, 315, 316
Vinyldene chloride, 167
Vinyls, 167
Visco-elasticity, 161, 173, 177, 180
Viscometer, Saybolt, 328, 329
Viscosity
  of oils, 261–263
  testing of lubricants, 328
Visible light and electronic behavior, 72
Visual examination tests, 318–320
Vitrification, 196, 201
Voigt model, 161, 162, 173, 177, 180
Volatile matter, 270
Volatility of gas turbine fuels and gasolines, 276
Volatility test, 335
Vulcanization, 179
Vulcanized rubber, 157, 163

W

Warping in lumber, 222, 223
Water, ionization of, 69
Water and sediment test, 334, 335
Water-cement ratio, 209
Water gas, 268
Waterproofing, 244
Wax crystals in oil, 331
Weaving, 243
Wedge grip in tensile testing, 290
Welding, 138, 150, 151, 154, 297, 321
Wet cell battery, 17
White cast iron, 81, 82
Wire, rolling of, 93
Wood
  adhesives for, 234
  defects in, 226, 227, 229
  industrial applications of, 235–239
  machine operations on, 233–235
  modifications of, 231–232
  properties of, 220–226
  standards and classifications of, 226–232
  strength properties of, 360–363
  structure of, 59, 217–220
  test specimens of, 294, 296
  types of joints in, 234
  use of, as fuel, 269, 271
Wood plastic composition, 231
Wood rays, 218
Wool, 242
Work hardening, 65
Wound continuous filament, 253
Woven fabrics, 243, 252

X

Xeroradiography, 323
X-ray as source of radiation, 321
X-ray diffraction analysis, 43

Y

Yarns, 243
Yielding during testing, 296
Yield point
  compressive behavior and, 12
  definition of, 62
  determination of, 285, 286
  failure and, 66
  speed of testing and, 292
  tensile behavior and, 12
Yield strength, 11, 286
Young's Modulus, 60, 224

Z

Zamak, 121
Zinc, 71, 117, 121
Zirconium, 125
Zyglo, 319